Critical Soc ᵤᵤₑₛ

Editors: JOCK YOUNG and PAUL WALTON

The contemporary world projects a perplexing picture of political, social and economic upheaval. In these challenging times the conventional wisdoms of orthodox social thought whether it be sociology, economics or cultural studies become inadequate. This series focuses on this intellectual crisis, selecting authors whose work seeks to transcend the limitations of conventional discourse. Its tone is scholarly rather than polemical, in the belief that significant theoretical work is needed to clear the way for a genuine transformation of the existing social order.

Because of this, the series will relate closely to recent developments in social thought, particularly to critical theory and neo-Marxism – the emerging European tradition. In terms of specific topics, key pivotal areas of debate will be selected, for example mass culture, inflation, problems of sexuality and the family, the nature of the capitalist state, natural science and ideology. The scope of analysis will be broad: the series will attempt to break the existing arbitrary divisions between the social-studies disciplines. Its aim is to provide a platform for critical social thought (at a level quite accessible to students) to enter into the major theoretical controversies of the decade.

The Radicalisation
of Science

Ideology of / in the
Natural Sciences

Edited by

Hilary Rose

and

Steven Rose

First published 1976 by
THE MACMILLAN PRESS LTD
London and Basingstoke
Associated companies in New York Dublin
Melbourne Johannesburg and Madras

ISBN 0 333 21140 5 (hard cover)
0 333 21141 3 (paper cover)
Typeset in Great Britain by Reproduction Drawings Ltd

Printed in England.

To the heroic peoples of Indochina, who
demonstrated to the world how to struggle successfully
against the science and technology of profit and
oppression

Contents

Acknowledgements

The history of these two volumes: *The Political Economy of Science* and *The Radicalisation of Science* with their common sub-title, *Ideology of/in the Natural Sciences*, is told in the Introduction which is reproduced in both books. Here, as Editors, we should like only to express our thanks to those who have helped in the creation of this collective endeavour. We must begin by thanking the contributors as individuals, who have not only given of their time and work freely in their own writing but also in their criticism and discussion of drafts of one anothers' chapters. Many of the chapters have benefited from being discussed at meetings within the polytechnics and universities and of trade unions and community groups. For those of us who work in the education or research systems the rediscovery of the sociology of scientific knowledge has provided a stimulating and critical environment for discussion of the concerns of these books. But it is to the radical science movement that our debts are greatest. To cite all the groups and individuals who have criticised and unstintingly aided the development of this work would be to give almost a history of the radical science movement. However, we should mention in particular: in Britain, *Science for People*, the Women and Science collective, *Radical Science Journal*, the Campaign on Racism, IQ and the Class Society, *Radical Philosophy* and the Indochina Solidarity Conference; in France, *Impascience*; in Italy the Science for Vietnam collectives; and in the United States the Science for the People groups in Minneapolis, Chicago, Boston and New York.

Many people have helped by reading and commenting on particular chapters, and we would here particularly mention: Francis Aprahamian, Pat Bateson, Paul Emmerson, Dot Griffiths, John Hambley, Altheia Jones, John Marriott, Ralph Miliband, Charles Posner, Jerry Ravetz, Ken Richardson, Fred Steward, Richard Whitley and Ron Wilson. Others have provided political and intellectual support over the period of production of the books — sometimes more than they may themselves have realised — Mary Evans, Mike Faulkner, Nora Frontali, Luke Hodgkin, Ian Muldoon, Helga Novotny, Felix Pirani, Esther Saraga, Tim Shallice, Joe Schwartz, Paul Walton, John Westergaard, Maurice Wilkins,

Charlotte Wolfers and John Wolfers. The help received from other friends and colleagues is acknowledged in various footnotes in the text. Janet Hackett, Liz Hainstock and Mac Foxley typed and copied many of the drafts; Laurie Melton chased endless references. Paulette Hutchinson translated the chapter by Jean-Marc Lévy-Leblond, 'Ideology of/in Contemporary Physics', and E. Maxwell Arnott translated Liliane Stéhelin's 'Sciences, Women and Ideology'.

We would also like to thank those editors and publishers who gave permission for already published material to be used. The publishing history is as follows: An earlier version of 'The Incorporation of Science' appeared in *Annals of the New York Academy of Science*, 260 (1975) pp. 7 – 21. 'The Production of Science in Advanced Capitalist Society' is to be expanded to book-form and published, in Italy, in 1976 by Feltrinelli. 'On the Class Character of Science and Scientists' first appeared in *Temps Modernes*, 29 (1974) pp. 1159 – 77. 'Contradictions of Science and Technology in the Productive Process' is based on earlier lectures at AUEW – TASS summer school in July 1972, the British Computer Society's Social Responsibility Group in November 1972, the Oxford Society for Social Responsibility in Science, March 1973 and the Conference *Is there a Socialist Science?* February, 1975. Earlier versions of 'The Politics of Neurobiology' were published as 'Do not adjust your mind, there is a fault in reality', in *Social Processes of Scientific Development*, ed R. Whitley (London: Routledge & Kegan Paul, 1974) and *Cognition*, 3 (1974) pp. 479 – 502, 'Scientific Racism and Ideology' draws on S. Rose, J. Hambley and J. Haywood, 'Science, Racism and Ideology', in *The Socialist Register* (London: Merlin Press, 1973) and also S. Rose, 'Scientific Racism and Ideology', in *Racial Variation in Man*, ed. F. J. Ebling (London: Blackwells for the Institute of Biology, 1975). An earlier version of 'Womens Liberation: Reproduction and the Technological Fix' appeared in *Sexual Divisions and Society*, ed. D. Barker and S. Allen (London: Tavistock Press, 1976). 'A Critique of Political Ecology' originally appeared in *Kursbuch*, 33. The translation is by Stuart Hood and appeared in *New Left Review*, 84, (1974) pp. 3 – 31. We are grateful to the translator and the editors of *New Left Review* for permission to reproduce it here. 'Science, Technology and Black Liberation' is revised from an original article which appeared in *The Black Scholar*, 5 (1974) pp. 2 – 8. 'History and Human Values: a Chinese Perspective for World Science and Technology' was originally a lecture for the Canadian Association of Asian Studies, Montreal, May, 1975, published in *Centennial Review*, 20 (1) (1976) pp. 1 - 35. We thank the Editor for permission to reproduce it, with minor editorial changes, here. 'The Radicalisation of Science' first appeared in *The Socialist Register* (London: Merlin Press, 1972) and was subsequently reprinted in *Science for People*, 21 and 22 (1974).

Introduction

In 1971, we began to discuss the idea of collecting together material for books on the theme of ideology of/in the natural sciences with other activists in the radical science movement. The response was positive and unequivocal. We could see that the political struggles in which the movement was engaged, beginning in different ways in the various advanced capitalist countries – the Indochina war and pollution in the United States, Britain, Japan and Australia; the hierarchy and elite nature of scientific practice in France and Italy – nevertheless were moving towards a series of fundamental questions which underlay them all. Scientists who had begun by feeling that 'their' science had been betrayed in the defoliation campaign in Vietnam, or that 'their' scientific community was a hollow myth, began to ask such questions as: Whose science is it? Who pays for it? Who decides it? Who benefits from it?

Because the production system of science requires interaction between workers at the international level, through journals, conferences, research centres, and so forth, concerns and issues which were felt in one section of the system rapidly spread and were taken up elsewhere. (In practice the movement had to learn that the vaunted internationalism of science was a function of its mode of production, just as much as contemporary capitalism demands the existence of the multinational corporation.) None the less, the differing political traditions – Marxist in France and Italy, social democrat in Britain and populist in the United States – meant that problems were seen and articulated in different ways. In France and Italy following 1968 there were laboratory occupations and attempts to develop self-managing scientific collectives involving the workers in particular institutes. In Britain, the campaign against chemical and biological warfare developed in pressure-group style, with attempts to use the media, ask parliamentary questions, persuade trade-union branches to pass resolutions, and urge moral renunciation of CBW work on individual scientists. In the United States, the work of the Honeywell collective centred on raising the consciousness of workers at Honeywell plants, designing and making fragmentation weapons for use in Vietnam, concerning the

nature of their product. In Japan, the campaign around mercury poisoning at Minamata involved grass-root mobilisation amongst communities
directly at risk from the pollutant. Some of these struggles were more
politically advanced than others, and even in any given situation there
were confused and contradictory ideas as to both the over-all strategy
of the campaign and the immediate tactics involved. Political action has
taken place in many different areas: within the scientific occupation
itself; in conjunction with factory workers; with local communities;
and in support of liberation struggles.

However, particularly in the United States and Britain, countries
with the most developed scientific production systems, and hence the
most organised scientific movements, these movements have been slow
to develop a theoretical perspective which would enable them to articu-
late the links between struggles in the different areas. While particular
groups have focused on, for instance, issues of the computer invasion of
privacy, or alternative technologies, there was little clarity about the
goals of the movement: was it to secure international law on CBW, to
unionise or to radicalise scientists, to aid workers in their struggle against
pollution in the work-place, or to act as a focal point in the general
struggle to overthrow the capitalist system? Instead, a cheerful and
energetic eclecticism prevailed. Initially, this was a strength, as new
spaces for action were found — spaces, it is important to say, that were
deemed not to exist by the old left orthodoxy — but by the early 1970s
most activists were recognising the practical and urgent need for theory.
They recognised that it was time to move beyond the early pragmatic
phase to a stage at which the contradictions present within science
could be seen as part of a general revolutionary perspective. This meant
not only strengthening the movement's understanding of its own strategy, but also delineating its enemies in class terms, for without this the
pragmatic eclecticism threatened merely to refresh and renew the
existing social order. Providing the enemy used the same language of
moral concern — and sometimes even the same populist rhetoric — it was
difficult to distinguish friend from foe. For instance, when the Club of
Rome — composed of those same industrialists and scientific elite who
had been in charge of the production/pollution process — announced
their collective concern with the finite earth, then, lacking a class perspective, the movement seemed only to carp ungenerously at a deathbed
repentance. The invocation of 'the scientific community', like that of
'the national interest', sought to bind strata with an antagonistic
relationship into one ideological whole.

The magnitude of the theoretical tasks confronting the movement — the need for a political economy of science in contemporary capitalism, its changing mode of production, the proletarianisation of scientific workers, the question of natural science as a generator of ideology, and of the ideology of science with its devaluation of all non-'scientific' knowledge, its elitism and the subtleties of its particular form of sexism and racism — all these needed definition and welding together theoretically. We had to achieve these tasks in the knowledge of the past history of theory and practice on the question of science in the revolutionary Marxist movement — and, in particular, the experience of the Soviet Union and China. Such an agenda was daunting for each of us individually — yet we all believed that to tackle it on the basis of our own separate experiences in different capitalist countries was imperative, and that collectively we could make a start. Geographical distance between us has meant that this has not been a fully collective programme in the sense that all its authors have participated in the writing of all sections, but rather that each has taken a particular section of the agenda and developed an analysis within a general shared framework, whilst in a few cases we have used material which was not written specifically for these books but seemed clearly in accord with their over-all theoretical position. By common consent, all royalties from the publication of the collection, in the several languages in which they are to appear, will go towards the development of scientific and technological education and reconstruction in Vietnam, by way of the Institute for Science and Technology in what was once Saigon, and is now Ho Chi Minh City, part of our recognition of the imperishable role that the struggle and sacrifice of the Vietnamese people has played in the theory and practice of revolution and of the transformation and recreation of human society.

The Political Economy of Science

The collection of essays have been organised into two volumes, with the common theme of ideology in/of the natural sciences. Whilst the two books are separate entities they reflect certain common concerns and are interrelated by a logical thread which this Introduction traces. The starting point has been an attempt to transcend both our own particular political pasts, and that of the revolutionary movement, which for too long has seemed to be polarised between, on the one hand, 'orthodox Marxism' with a rigid belief in the objectivity of the natural sciences as a model to which Marxism, as scientific socialism, aspired, and on the other, an anarchism which has seen scientific rationality itself as part of the enemy. In order to recreate a revolutionary critique of the actual social functions of science as they exist in today's capitalist and state socialist societies, it is necessary to understand the origins and the limitations of this 'orthodox Marxist' view of science, which regards itself as operating in a tradition which stretches from the most recent pronouncements of the Soviet Academy of Sciences back through Stalin and Lenin to Engels, and hence Marx himself. We therefore begin, in the first chapter of *The Political Economy of Science*, by returning directly to what Marx and his close collaborator Engels themselves wrote about science, and in doing so rediscover in Marx those compelling theoretical insights which, however briefly and schematically they are presented, lie at the core of every one of those questions of theory and practice which are the concern of today's movement.

The second chapter of *The Political Economy of Science* moves directly forward from Marx and Engels to the issues of the 1970s, with which the whole of the rest of the book is concerned. In 'The Incorporation of Science', we ask what features characterise the present social function of science in Western capitalist societies and the Soviet Union. We argue that, today, science has two major functions, as part of the systems of production and of social control. Especially since the Second World War, science has itself become industrialised and enmeshed in the machinery of state. We examine two myths, the liberal academic myth of the autonomy of science and the 'orthodox Marxist' belief in the inevitable contradiction between science and capitalism, and show that neither accounts for the actual development of science and 'science policy' — the management of science — as it has occurred in Britain or the United States. Faced with capitalism's fusion of science and oppression, and the conspicuous failure of the Soviet Union to avoid the same development, the 'Frankfurt School', typified by such writers as Habermas, has claimed that scientific rationality is *inevitably* oppressive and has abandoned that optimism with which Marxists had maintained the automatically progressive nature of science. The question is whether capitalist science represents an unavoidable and fatal attempt at the domination of nature, or whether it can be confronted as a 'paper tiger', to make way for a genuine science for the people.

The next four chapters discuss in greater depth the questions, raised in Chapter 2, of the role of science in production and the consequences for scientific workers. This issue raises fundamental questions for Marxists both at the theoretical level and in terms of political and organisational strategy. In the first place, where does science fall within the Marxist categories of 'base' and 'superstructure'? Is it part of the productive process? This is not an abstract question, for if it is purely superstructural, then scientists, whatever the contradictions within their role, cannot be regarded as workers, but primarily as within or associated with the ruling class, either by assisting in the structural maintenance of the capitalist apparatus, like lawyers or accountants, or as transmitters of its ideological values, like teachers or journalists. That is, they will in general find that the contradictions of capitalist society do not oppress them but serve to protect their privileges and position. On the other hand, if science is part of the productive process, 'scientists' are really scientific *workers* who sell their labour to the capitalist in parallel with other workers; like other workers, they become alienated from their creations, from the products of their labour – in a word, they are proletarians, and as such form part of the potential revolutionary forces within society.

This issue has long been a source of debate and discussion because upon it hangs the question of whether, politically, scientists are to be seen as friend or foe. This is particularly important in the present period of the incorporation of science, and the answers given by Marxists in earlier periods may no longer be appropriate today. These chapters argue, essentially, that science spans *both* base and superstructure; it has both a productive and an ideological role, the understanding of which is confused by reference to 'the scientific community' as an undifferentiated whole. In fact, this 'community' is divided into, on the one part, the majority of alienated, proletarianised *scientific workers*, and, on the other, the tiny majority of the elite carriers of bourgeois ideology, the *scientists*.

Chapter 3 is by a group of physicists and mathematicians, Giovanni Ciccotti, Marcello Cini and Michelangelo De Maria, associated with the *Manifesto* group in Italy. They approach the question of the role of science as a productive force from the perspective of Marx's theory of value. Today, they conclude, the role of applied science and technology can be seen as the production of information as a commodity, to be sold on the market just as are material commodities. The relation of scientific workers to their product is therefore comparable to that of manual workers; they are alienated from it. Science as commodity production is thus the dominant mode, which serves as a model for the style of work even in fields which are not directly concerned with the production of information for sale, such 'pure' sciences as high-energy physics or biology. These fields have a dual role, generating an information 'base' on which the information-commodity market can rest, and serving as test-beds for the checking of advanced technology.

Chapters 4 and 5 take up the consequences of this role of science as a productive force for scientific workers themselves. André Gorz, the editor of *Temps Modernes*, asks: what are the implications of describing scientific workers as proletarianised? Science is still a privileged, elite activity: in industry scientific methods may be used by some categories of workers (production engineers for example) to oppress others by means of speed-ups and other forms of technological rationalisation; none the less, the fragmentation of scientific knowledge, and its ideological values, has come to make intellectuals increasingly the victims rather than the beneficiaries of the class system. The way forward lies in ridding expertise of its class nature, of breaking the barrier between expert and non-expert.

To a large extent, Mike Cooley shares Gorz's preoccupations, but brings to them the perspective of the shop-floor struggles which his own designers' and draughtsmen's union (TASS, a section of the Amalgamated Union of Engineering Workers, AUEW) has been involved in. Cooley shows how the increasing cost and rapidity of obsolescence of fixed capital impose increasing demands on both manual and intellectual workers in industry, with speed-ups, shift work, fragmentation of skills and dehumanisation. This proletarianisation began, as Gorz points out, in the chemical industry in the nineteenth century, but has now spread to designers and draughtsmen, architects, computer programmers and mathematicians in industry. However, as Cooley shows, a capitalism based on very complex, very expensive technology, develops the weaknesses of its own strengths. It is these points of vulnerability which proletarianised scientific workers, side by side with their manual worker comrades, must learn to probe and enlarge if the system is to be shattered and social transformation to occur.

The remaining chapters of *The Political Economy of Science* are concerned with a distinct theme, whose roots, as we show in Chapter 1, derive from Marx's and Engels' own writings, but which has burgeoned into major significance in recent years. This is the theme of the struggle between ideology and science within the natural sciences themselves. The analysis of this struggle is no easy task. Ideology is of its nature mystifying. Where the sharpness of the contradictions within the capitalist mode of production continually force themselves into the consciousness of the worker, the very role of ideology is to obscure these contradictions and diminish the level of consciousness. Hence, whilst the superstructural battle and that in the work-place are part of the same conflict — indeed, they continuously interact — the dominant class pretends that there is no ideology, and so no grounds for battle: that science has once and for all driven out all ideology. In the second place, because of the abortive nature of the Soviet cultural revolution and the experience of Lysenkoism (discussed in Chapter 2 of *The Political Economy of Science* and Chapter 2 of *The Radicalisation of Science*), the continuity of the critique of ideology has been ruptured. Marxists are faced not only with the problem of starting afresh from the moment of rupture, but also with the analysis of the rupture itself. For many years, orthodox Marxism in its preoccupation with the objective world laid to one side complex questions of the superstructure, arguing for the most part that it was determined by the economic base; natural science, while belonging to both, was above ideology.

Yet battles in the superstructure are not some revolutionary luxury item which can be dealt with after the workers have destroyed capitalism, but are intrinsic to the political struggle itself. No one writing in these books has gone out to look for 'ideology in astrophysics', 'ideology in inorganic chemistry', in cell biology, biochemistry, and so on in the way which it seems Marxist scientists did in the 1930s, clutching their *Dialectics of Nature* and searching for thesis, antithesis and synthesis in the particular bit of the natural world they worked in. Instead, work on science's role in perpetuating racism, exposing the implications of reproduction science for women, or the nature of the politics of ecology, has been written as part of an on-going struggle, not as an item of an academic agenda. For this reason these chapters do not represent an even spread over the natural sciences. So long as most of the current struggles relate to the biological sciences, then it is right that we work in this area. (It is not however the case that the cultural analysis in some sense 'follows' the existence of struggle at the point of production, nor is it a question of awarding prizes for priority in discovering racism to the Mansfield hosiery workers or those working on scientific racism, but rather that each should see the other as necessary.)

Chapters 6 and 7 of *The Political Economy of Science* interlock, in that the second, on scientific racism, is a special case of the critique of ideology in the neurobiological sciences contained in the first. Both chapters argue that many of the theories and linked technologies of neurobiology, from drug therapy through behaviour modification to IQ testing, are fundamentally biologistic. Biologism takes one part of the explanation of the human condition, excludes all other considerations, and announces that it has *the* explanation for aggression and altruism, war and class struggle, love and hate. Attempting to change the human condition is then presented as an absurd opposition to both our natural selves and the natural world. The everyday possibility and actuality that men and women have continuously changed their situations in the course of history is methodologically and philosophically excluded. Biologism, for all its apparent scientificity, is thus mere ideology, the legitimation of the *status quo*. It is a method not of explaining people, but explaining them away as 'nothing but' assemblages of molecules, larger rats, naked apes or hairy computers. In biologism, reductionism, which was originally simply a powerful tool for examining specific problems under rigorously defined conditions, becomes saturated with ideology. Reductionism is thus part of the ideology *of* science, and in so far as the theories serve specific dominant classes, also legitimises and obscures ideology *within* science. The particular importance of biologism derives from the nature of the fight in which the bourgeois state must presently engage to protect itself. Where in the past its military effort was primarily against other nation-states or directed towards securing new colonies, with internal control a related but subsidiary question, since the growth of revolutionary guerilla movements, the main enemy is within. Faced with this internal enemy, methods of social control become of paramount importance to capitalism; biologism with its ideological justification and its techniques of manipulating and controlling people comes to the rescue.

Chapter 8 of *The Political Economy of Science*, while still concerned with biology, sets out to analyse the ingrained sexism of current developments in reproduction technology, from genetic engineering to hormone time capsules. This characterisation of science is opposed to that of the radical feminists such as Shulamith Firestone who see technology as essentially neutral and therefore capable of generating a 'technological fix' for the reproductive role of women. By contrast, the chapter argues the need to link the class and the women's struggle in the pursuit of human liberation, where science would serve the goal of nature humanised, and 'the long struggle from nature to a truly human culture' would be advanced.

The final chapter, by Hans Magnus Enzensberger, West German poet and political activist, is a critique of political ecology. In it, Enzensberger is concerned with two tasks. One is to expose the ideological role played by the prophets of the ecology movement as it has mushroomed since the late 1960s, people like the Ehrlichs, Forrester and Meadows, the MIT modellers of 'the limits to growth' and the 'Club of Rome'. Enzensberger lays bare the links between the 'ecology movement' and imperialism, and shows that in their frequent apocalyptic pronouncements, the doomsters are playing a deeply ideological role. The second point is that the concern over pollution or global destruction cannot be dismissed as pure ideology or merely a consequence of capitalism that the transition to socialism will automatically resolve, as some Marxist groups tend to argue; this itself becomes an ideology which ignores the real material base for much of the present concern. The ecological hazards are not to be dismissed as trivial, and even after the destruction of capitalism they will remain major problems. 'Socialism, which was once a promise of liberation, has become a question of survival. If the ecological equilibrium is broken, then the rule of freedom will be further off than ever.'

The Radicalisation of Science

Whilst *The Political Economy of Science* is concerned primarily with the critique of existing capitalist science, much of the discussion in *The Radicalisation of Science* deals with attempts at its transformation. The first chapter of the book, originally written for the 1972 issue of *The Socialist Register* and subsequently reprinted in *Science for People*, gives the book its title. It represented the gathering together of our personal experiences within the scientists' movement at that time, an attempt to describe the origins, the brief history and perspectives for action of the movement. Even though our understanding of certain of the issues has sharpened in the intervening period, we decided to reprint it as it stands, both because it has served to fuel a necessary debate within the movement in the last few years, and because it represented the original programmatic guide for the present collection. However, we have updated it, and added a postscript from the vantage point of 1976. Chapter 2 takes up a topic which no discussion of the relationships between Marxism and the natural sciences can avoid. This is the Lysenko 'affair', 'problem', 'scandal' — as it has variously been described. Coming at a crucial time in the development both of the Soviet Union and of the attempts by Marxist scientists in the West to grapple with the problem of the relationship between science and social structures, it seemed to provide the acid test of the possibilities of a socialist science. The consequences of the debate were disastrous — concretely for the geneticists who lost their lives in Stalin's camps, for the development of Soviet genetics (and, less certainly, Soviet agriculture) and theoretically for the very idea of a socialist science. The period following 1948, the high point of Lysenkoism, marked a retreat in the Soviet Union to a 'neutral ideology of science, and, in the West, a turning away of many scientists from the orthodox communist parties and even from Marxism itself; they were 'forced to choose between their science and their political convictions'. As the period of Lysenkoism retreats, so it gains a mythology, and even Marxists have shied away from attempting to peel off these mythical accretions so as to subject the episode itself to rigorous Marxist analysis. Yet it is essential that we understand what happened, it only to help avoid a repetition of old mistakes. As Richard Lewontin and Richard Levins make clear, it is no good merely to see the episode as an example of the workings out of the 'cult of the personality', or a dreadful warning of the consequences of mixing biology and politics — nor yet as the high point of Soviet science before its retreat with the rise of revisionism. Rather, we must seek its roots in the objective conditions of Soviet agriculture and society, and understand it as an aspect of the tentative and inadequately articulated attempts within the Soviet Union of the 1930s and 1940s to achieve a cultural revolution — but one monstrously distorted by its imposition 'from above' by a mixture of administrative fiat and terror, rather than 'from below' by a creative social and political upsurge amongst the people themselves.

What was Lysenkoism most directly about, and what were its claims? As mathematical biologists, whose own research relates directly to the substance of the Lysenkoist claims, and as themselves politically engaged within the Science for the People movement in the United States (both refused membership of the US National Academy of Sciences on the grounds of its involvement with the Department of Defense and its perpetuation of the hierarchical, elite structure of American science) Lewontin and Levins are well placed to make the assessment. They begin by assessing the present significance and interest of the Lysenkoist controversy. They then briefly summarise the philosophical and scientific claims of Lysenkoism itself: what were Lysenko's views on heredity and its relationship with the environment? (It might be helpful to those unfamiliar with the details to compare this discussion with that in Chapter 8 of *The Political Economy of Science*, where some of the same issues are discussed in relation to the IQ debate.)

Lysenko's views are contrasted with some of the almost mystical concepts which many classical geneticists of the Weismann school at the time held about the gene and its relationship to the environment. Then, in a crucial section of the argument, they discuss the objective conditions creating Lysenkoism: the weakness of Russian agriculture and its climatic problems, and the implications that these latter had for the interpretation of experiments and the use of statistics. The weaknesses of existing genetic theory, and its ideological role and links with philosophical reductionism and racism, are analysed.

Other vital factors were the reaction of the Russian peasantry to collectivisation, and the elite, bourgeois structure of Russian science which still remained the case even twenty years after the 1917 revolution. It is this feature – the challenge to the bourgeois expert – which represented that part of Lysenkoism which can be seen today, with the hindsight provided by the Chinese experience, as the attempt at cultural revolution.

Lewontin and Levins conclude by asking: Can there be a Marxist science? The answers they give, in terms of what the dialectical method can and should mean in science, may serve, in their emphasis on the unity of structure and process, the wholeness of things and the interpenetration of an object and its surroundings, as a key and summary statement of the major themes of both books.

The next two chapters of *The Radicalisation of Science* are, concerned with the nature of the institution(s) of science as they have developed under contemporary capitalism, and particularly its sexist character. Monique Couture-Cherki, a solid-state physicist from Paris, and Liliane Stéhelin, a sociologist of science from Strasbourg, raise the question of sexism. Couture-Cherki points to the systematic exclusion of women from the higher ranks of science, their concentration in subordinate positions, and the powerful ideological pressures which are exerted to systematically exclude women from scientific achievement. Amongst these, the most powerful are the ideology of the family and the persistent attribution to women of more 'docile', 'feminine' characteristics, 'not appropriate to high scientific achievement', and so on. But can these be overcome? Liliane Stéhelin takes this question as her starting point. For her, the present forms of science are fundamentally interlocked with sexist, male ideology. In order to succeed in science, a woman is required to submerge – overcome – her feminine character and become an honorary male. To do this is the ultimate trap. Indeed, we can expect, at least in periods of labour shortage and capitalist expansion, to see a steady effort made to eliminate the obvious barriers to women's progress in science, the provision of creches and better maternity arrangements, more efforts at 'equal opportunity' appointments, and so forth – if only because women represent a reserve of productive forces.

Yet the production code of science, its ideology, will remain fundamentally masculine; forced to compete within it, women will either succeed by denying their femaleness, or fail, confirming their inferiority. The task, therefore, is the attack on and subversion of the masculine code itself, which raises the question of whether there is indeed a feminine science as an alternative to masculine science in the same way as there is a socialist as opposed to a bourgeois science. This question leads Stéhelin into a consideration of the social and psychoanalytic view of women and into the question of the resynthesis of Marxism and psychoanalysis which has been a major concern of French Marxism in recent years. Can the masculine code of science be overcome? If so, she concludes, there is 'the promise that one day other women (with other men?) will be able to open the way for a new science'.

It is against this background that it becomes possible to raise the question of just what can be learned from the Chinese experience. Despite the greater accessibility of China, and the enthusiasm for what are seen as the lessons of the cultural revolution, an adequate account of what has been and is being achieved in China must start from an understanding of the particular circumstances of China's own social and economic development, rather than from timeless universals. Joseph Needham's chapter was originally given, in 1975, as a lecture in Montreal, and its lecture form is preserved here. In it, he first describes his own history and that of the *Science and Civilisation in China* project (Cambridge University Press, 1954 onwards), and then sets out to counterpose the historical development of Chinese science with the contradictions of science and the anti-science movement in Western capitalism as typified by, for instance, Theodore Roszak. Needham argues that the anti-science movement has emerged in the West in response both to the social function of science under capitalism and the claim that science represents the only valid way of understanding and apprehending the universe – an aspect of the scientistic ideology of science with its overriding aim of the domination of nature. By contrast, he shows, the Chinese have historically never had such a scientistic approach nor fallen prey to reductionism. This is not to say, Needham emphasises, that the practice of science in today's China has nothing in

common with that under capitalism, but it is a practice reflective of a
dialectical conception of the interrelations of nature and humanity, and
of a science done for and with the participation of the people as a
whole. Needham's analysis is couched in characteristically more ethical
and religious language than is familiar to many activists in the radical
movement today; a language from within the tradition of English Christian communism, its moral passion echoing that of Digger Winstanley.

Chapter 6 is derived from an article in *The Black Scholar*, 'Science,
Technology and Black Liberation', by Sam Anderson, a New York
mathematician. In it, Anderson briefly outlines some reasons for the
technological underdevelopment of Africa by European colonialism and
the role of science in the emergence of capitalism, leading to the present
situation in which, for the Third World countries, science has the two
aspects of 'liberation' and 'exploitation'. The position of the black
scientist in the United States (or Western Europe) has much in common
with that of the woman scientist discussed by Couture-Cherki and
Stéhelin — forced into an alien, bourgeoisified role. To combat this,
and to contribute needed scientific and technological skills for the
movement, Anderson calls for black scientists to organise.

The final chapter, by Jean-Marc Lévy-Leblond, theoretical physicist
and one of the collective producing the radical science magazine
Impascience, spans the themes of both ideologies, *of* and *in*. Because
modern physics is a discipline founded at the birth of capitalism, it is,
in certain important respects, the model to which all science aspires.
Although its theories may have little ideological significance in themselves, physics as a social and cognitive institution is saturated with
capitalist ideology, and the ideology of physics as a science becomes the
dominant theme of Lévy-Leblond's chapter. To mathematise, to forma-
lise, becomes the hallmark of the mature, hard science against the
immature, soft science (the masculinity/femininity — superior/inferior
metaphor is not lost). Nor is this only an issue in the natural sciences, as
physics becomes the model for all human knowledge, and what cannot
be encompassed by its mode of rationality is illegitimate.

Physics is thus at the heart of the ideology of expertise: the claim
that, to be a physicist, particularly a theoretical physicist, gives an
individual *as of right* the power and knowledge to speak with competence in almost any area.

Within physics, social practice is deeply hierarchical between scientist and student or technician — symbolised by science's reward system,
at the peak of which come the Nobel Prizes. The Laureate, in fact a

narrow specialist, becomes transmuted by social alchemy into one of Plato's Men of Gold, to whom all humanity must defer. Another aspect of the hierarchy though is the divorce between theory (high prestige) and practice (low prestige), epitomised by the elite nature of theoretical physics and the lower status of the experimental science of engineering. Lower still, yet equally hierarchised, comes teaching. This divorce affects the development of the subject of physics and, at the same time, lays it open to the type of ideological exploitation discussed in relation to biology in other chapters. The divorce from practice means that physicists are increasingly concerned with an artifical world of their own construction, outside the experience of common problems which physics used to be concerned to explain. The solution for these problems will be the solution for science as a whole.

*****

The themes of the chapters in these two books reflect a common agenda, an agenda shared with many of the activists in the radical science movement who have been discussing and working out these issues in practice over the last few years. At an earlier stage, many of the chapters have formed part of, and been improved by, this discussion. By collecting and developing the arguments on paper, we believe that the theory and practice of the movement will be advanced. Nevertheless, it is important not to forget differences. These reflect the fact that we belong to a social movement with diffuse aims and not to a single party with a clear line and agreed priorities. What we hold in common is a desire to work towards a new society where a new science and technology can serve the interests of all the people.

1

The Radicalisation of Science*

Hilary Rose and Steven Rose

THEMES

Since the end of the 1960s there has been a clear shift in consciousness of many scientists — especially science students — of the role of science and technology in contemporary capitalism. This movement has been concentrated in the United States and Britain, the two most scientifically advanced Western countries, judged by such formal criteria as percentage of GNP spent on science, or numbers of papers published or Nobel Prizes per head of the population.

Significant developments however have also occurred in Belgium, France, Italy and West Germany. The movement has embraced scientists and non-scientists alike in a variety of shifting and uneasy alliances of old and new left, liberal concern over actual or potential scientific abuses, explicitly Marxist attempts to analyse the contemporary functions of science and technology under capitalism, and libertarian, anti-scientific, or even frankly reactionary attacks on the rationality of scientific method and the anti-human quality of technology. The movement is still in rapid development and has few obvious historical parallels. This chapter attempts an analysis of its directions and future

---

*This chapter was originally written for the 1972 issue of the *Socialist Register* and subsequently reprinted in *Science for the People*. It represented the gathering together of our experiences within the scientists' movement at that time. We reprint it here — even though our understanding of certain of the issues has sharpened during the intervening period — because it served both to fuel the necessary debate within the movement concerning its direction and theoretical needs, and also as a programmatic guide for the present book. Gary Werskey's critical appraisal in *Radical Science Journal*, 2/3 (1975) must be mentioned in this regard. We have taken out or compressed some material which is covered elsewhere within the book or now seems merely past history, added some footnotes and a postscript.

perspectives. As actors as well as observers in the movement, our account will focus on those aspects of which we have most detailed knowledge and which seem to us most relevant; hence we mainly discuss Britain and to a lesser extent the United States and Western Europe. No reference to the interesting parallel developments in Japan has been made, an omission which reflects our own ignorance.

Three major themes have, in varying degrees, been reflected in most of the groupings and debates that have occurred in the past few years.

(1) *The Abuses of Science*

This is perhaps the broadest area, ranging from concern over environmental issues, such as non-returnable bottles, DDT in mothers' milk and penguin fat and the 'population explosion'; through fears of potential scientific advances such as genetic engineering, psycho-social control technologies and the computer invasion of privacy; to an explicit recognition of the harnessing of science to the purposes of imperialism, as in the application of new technologies to the war in Vietnam and the absorption of university science into secret research funded by defence departments. Although analyses of such abuses have often been politicised so as to demonstrate that they are an inevitable consequence of doing scientific work within an oppressive social order, many scientists have tended, none the less, to see them as indicative of external pressures upon a science regarded as a relatively autonomous (neutral) activity; the pressures generating, therefore, particular perversions or abuses of an otherwise desirable — even progressive — activity.[1] The dominant ideology of science, particularly in the elite sectors of university and research institutes, generates a mystifying conception of the scientist's work, stressing his ownership of his paper (and it almost invariably is 'his'), his membership of the community of science, such that 'uses' and 'abuses' are seen as independent and unrelated to his activity. It is the strength of this belief which leads to issues being posed — perhaps even necessarily in view of the very low level of consciousness — in terms of personal or even social responsibility, rather than in terms of alienation.

(2) *Science as a Non-Neutral Ideologically Laden Activity**

Here the attack on scientific abuses goes deeper to consider the nature of the scientific process itself. Science done within a particular social order, it claims, reflects the norms and ideology of that social order.

*See next page.

Science ceases to be seen as autonomous but instead as part of an interacting system in which internalised ideological assumptions help to determine the very experimental designs and theories of scientists themselves.[2]

One consequence of this analysis has been that attacks on the role of science under capitalism have become expanded into attacks on science and scientific method as themselves products of an oppressive capitalist order. As a consequence of a mystification which has elided the methods of science with its goals, a chorus of voices — ranging from the drop-out to the religious high conservative — has ascribed contemporary social problems in large measure to the rationality and scientific method which lie at the core of technology-based industries. A retreat from industrialisation towards a simpler life style, and a belief system based on anti-rational premises, typified by much of the environmentalist movement, are the proffered solutions.[3]

(3) *The Self-Management of Science*

This area of debate takes the argument within the laboratories themselves; how are research programmes and individual experiments planned and carried out? In whose names are they published? Scientific tradition is elitist, dominated by individualistic expertise, despite the ostensible communality of the scientific method. Can laboratory work itself be collectivised and democratised, the barriers within the laboratory and between the laboratory and the external world be broken down?[4]

With these themes in mind, we can now trace the roots and development of the present movement.

HISTORICAL ANTECEDENTS OF THE RADICALISATION OF SCIENTISTS

Recent developments in the radicalisation of science and scientists must be seen as the culmination of a lengthy historical process during which the scale and nature of the scientific endeavour have been transformed,

---

*In 1975, with much more work available on this question, it is easier to distinguish the two forms of ideology of/in science. The ideology *of* science is the hierarchy of knowledge and the hierarchy of the knowledgeable; ideology *in* science is exemplified by such themes as scientific racism.

and the work situation and consciousness of scientists consequently profoundly modified. The late 1960s saw the end of a period of virtually continuous scientific expansion in Britain stretching back about as far as the seventeenth century. This expansion, in terms of both money spent and scientific knowledge gained (at least defined as papers published), had represented a regular doubling every sixteen or twenty years.[5] From the nineteenth century onwards the reciprocal links between science, technology and society became steadily firmer and more binding, resulting through that century and to the First World War in a steady incorporation and institutionalisation of the activity and consequences of science. (See Chapters 2/3 in *The Political Economy of Science*.) At the same time this institutionalisation permitted − indeed, even encouraged − the powerful mythology of scientific autonomy by which research − especially university research − was seen as pure and divorced from technology, industry and all their doings. This ideology of autonomy was seriously shaken by the 1914 − 18 War when for the first time the technological weaknesses of the British Empire became so dramatically exposed that direct government intervention into the management of science, the establishment of the Department of Scientific and Industrial Research and the first joint state/private-enterprise co-operative research associations, became necessary.[6]

The inter-war years saw the slow response of scientists to the changed situation beginning with the early attempts at unionisation. The National Union of Scientific Workers, later called the Association of Scientific Workers', was established with an explicitly socialist platform in 1918. At this point the problems of science were seen, even amongst socialist scientists, as predominantly related to economics − under the circumstances of British capitalism between the wars, scientific spending was pitifully small; an expansion in the scale of scientific spending, coupled with the adoption of centralist planning techniques, could, it was argued, release the transforming powers of science, help it liberate the world. Meanwhile abuses which arose from particular applications of science represented, it was assumed, the consequences of attempting to conduct research under capitalism. By the 1930s the model for the harnessing of science to the service of society was seen, for most of this group of scientists, as the Soviet Union.

It was indeed from the Soviet Union that the second of our major themes, that of the ideological determination of science, was injected into the British debate with the appearance of the Soviet delegation at the 1931 London conference on the history of science.[7] Although the delegation was headed by Bukharin, its major contribution was provided by a paper from Hessen on *The Social and Economic roots of Newton's Principia*. This argued that the *Principia*, the high point of seventeenth-century science.was not an isolated product of scientific genius generated by the internal logic of science alone, but rather had emerged as a consequence of the needs of the developing British bourgeoisie. The theories presented by the Soviet contributors introduced at least one aspect of Marxist analysis of the internal content of science to an audience, partly of highly empiricist British historians of science, and more importantly to a group of young British left-wing scientists. Needham, who was among those present, has suggested, in the 'Preface' to a later reprinting of *Science at the Crossroads*, that even they did not immediately perceive the significance of the Hessen paper.

The Soviet contributions were derived from debates on what a socialist science would be, which had begun to develop during the late 1920s and the early 1930s in the Soviet Union. These debates had two strands: the first was drawn from the discussion in Engels' *Dialectics of Nature* and Lenin's *Materialism and Empirio-Criticism*, which had explored the relationships of dialectical materialism to scientific knowledge.[8] In these strands of the debate the objectivity of the natural sciences was assumed, and represented the model towards which Marxism, as a scientific socialism, should approach. As objective science proceeded, Engels assumed, dialectical laws would be found to operate in nature; Lenin's contribution was the argument that materialism demanded a copy-theory by which perception was related to reality. It was this first strand in the Soviet debate which was to become of major significance internally in the late 1930s and again in the period 1948 to 1953.

The second strand raised the question of whether a socialist society would generate a specifically *socialist* science; was there an unique socialist biology, by contrast with bourgeois biology, for instance? In so far as Newtonian mechanics were seen by Hessen as the product of a particular historical period in bourgeois society, the answer to that must have been seen as in the affirmative; what Hessen's contribution in 1931 (and indeed subsequent Soviet discussions in this area) have not adequately analysed out, however, is the question of whether there is indeed a bourgeois, by contrast to a socialist, science.

But the unravelling of this argument, though implicit in Hessen, was not perceived by the Marxist British scientists in the 1930s. Rather, like

Haldane, they were to spend their theoretical strength over the next few years in a relatively fruitless endeavour to demonstrate the negation of the negation, the interpenetration of opposites, and the transformation of quantity into quality in a variety of scientific developments.[9] Only when, much later, Needham turned his attention to the history of Chinese science and technology[10] and Bernal attempted first the seminal *Social Function of Science* (1939) and later the rather more synoptic and less satisfactory *Science in History*, was the Hessen experience to bear fruit.[11]

Meanwhile another strand of relevance to the later debate emerged with the foundation in 1932 of the Cambridge Scientists' Anti-War Movement, a loose grouping which was to provide the scientific and political cradle of many of those who were to become the eminent supporters of the first phase of the radicalisation of the 1960s. The group concentrated its attacks on the abuse of science by the military, and many of its members were to become associated with attempts to rescue Jewish scientists from Hitler Germany. Later they were to be involved in the big ARP campaign to improve civilian precautions against air attack.

But the apotheosis of the 1930s movement was certainly Bernal's *Social Function of Science*, which effectively summarised the critique of science under capitalism which the group had developed. For our purposes there is one aspect of the book which is strikingly in contrast with the present situation. Bernal's science is optimistic, of its nature progressive and hence the natural ally of socialism. Socialism above all means planning, the rational and logical exploitation of science for the benefit of the people.* Only plan, increase the scale of scientific spending some tenfold (to the level of that current in the Soviet Union for example), defeat capitalism and all would be well. The ideology of what was to become techno-economism began to take shape. The elitist assumptions of this model of a socialist science, which were certainly reflected in the work relations of the Marxist scientists themselves, were to represent a source of trouble that were only to begin to be analysed more satisfactorily in the late 1960s. In its early days the AScW had

---

*In retrospect this reading of Bernal seems unjust: he was well aware of the class structure within science and its bitterly hierarchical character, and that under capitalism science would serve capitalist ends. But for Bernal, immanent within capitalist science, with its exploitation and oppression, was a socialist science which would truly be a 'science for the people'. J. D. Bernal, *Marx and Science* (London: Lawrence & Wishart, 1952).

two grades of membership, one for scientists, the other for technicians, and there had been serious debates as to whether technicians should be admitted at all.

1939-1945 was to complete the process which 1914-1918 had commenced, the rationalisation and institutionalisation of the organisation of science itself, with the energetic support of even the most *laissez-faire* members of the scientific community. A register of scientific manpower was completed, and scientists permeated most branches of government. The anti-war group of a few years previously now advised on the most effective types of bombing patterns, radar, chemical and biological weapons, or defence against them, and of course, the Bomb. The Bomb above all meant the mobilisation of scientists; individual brilliance became secondary to team work in the biggest co-ordinated research and development effort ever. For many scientists the brief years of the New Atlantis at Los Alamos, before Hiroshima and Nagasaki, were to remain thereafter a creative and emotional high point in their lives. Yet the mushroom clouds of the two explosions symbolised, as all involved clearly recognised, the end of an era.

By the end of the war, the autonomy of science had become a myth helping to ensure that those students with first-class degrees remained in the universities and those with second classes went into industry. The utility of science to capitalism and imperialism had been demonstrated with a classical simplicity which was to result in the exponential growth of research and development budgets with scarcely a question from government or people for the following twenty years. In addition, however, the public image of science began to change from the ambivalent to the downright malevolent. Perhaps most important, the scientists' consciousness had been modified because they felt themselves not merely responsible for the production of the Bomb but its very conception, by proposing its manufacture to governments both in Britain and the United States whilst at the same time they had failed to prevent its use against Japan.

This conjecture of attitudes amongst scientists was to characterise their political style in subsequent years. Partly because the Bomb had been so devastating, the scientists, as an elite, seemed to assume that that the research they did, unlike that of the historian or artist, reacted that the research they did, unlike that of the historian or artists, reacted very directly upon society. At the same time, there was a danger that the Bomb would be seen as an *inevitable* result of physics, so that anyone with qualms concerning its use would have to stop doing physics;

the responsibility would have been too much (indeed many did go into biology for these reasons). The defence against this criticism was to claim the *neutrality* of science, as a force either for good or evil depending upon the whims of society. This convenient conjuncture enabled many scientists in the two decades that followed to continue simultaneously to do high science — even accepting research grants from the military to do it —while at the same time professing radical political attitudes or arguing against particular developments in the arms race; thus the discord between the objective reality of the uses of science and the consciousness of the scientists became almost complete.[12]

This move towards the attempted divorce of science from politics, leaving only the uses of science in dispute, was hastened by events in the Soviet Union. The debates over the distinctions to be drawn between socialist and bourgeois science which had begun in the late 1920s had long since been confounded with the vices of Stalinism; the equation of Russian patriotism with socialism, so that even bourgeois Russian scientists of the nineteenth and early twentieth centuries were to be lauded to the skies (Lobachevsky, Butlerov, Popov, Pavlov, Michurin); the ossification of official dialectical materialism; the imposition of bureaucratic solutions to intellectual problems; all combined with a desire to adopt scientific strategies which would most rapidly yield technological and industrial pay-off.

In almost every branch of fundamental science serious questioning of theoretical formulations became hopelessly and tragically confounded with a variety of these other factors. Some of these debates began in the 1930s, others were to appear only after 1945; all were to reach their maximum intensity over the period 1948–53, where they are often (wrongly it would seem) associated with the high peak of Zhdanovism in cultural control.[13] Complementarity, quantum theory and relativity in physics were attacked as undialectical and unmaterialist (at one point Einstein was under attack both in Nazi Germany and in the Soviet Union). The idealism and mysticism of Jeans and Eddington in astronomy in the 1930s and the later 'big bang' theories of the origins of the universe were challenged. In the late 1940s Linus Pauling's resonance theory in chemistry which claimed that for some molecules there exists no single unique structure, came under heavy assault as idealist.

But the weightiest challenge to Western scientific orthodoxy came in biology — the genetics dispute associated with Lysenko.[14] (See Chapter 2 of this book.) In principle, as Haldane and the American socialist geneticist Müller, who worked in the Soviet Union in the mid-1930s (and indeed many others) recognised, the concept of the gene as a chemical entity is thoroughly materialist. It can be isolated, it has a defined chemical composition, and indeed it can be manipulated. Müller, for example, put forward a programme of socialist eugenics, including the preservation of Lenin's sperm for subsequent implantation which would surely have appealed to Stalin in his most grandiose moments.[15] Soviet plant genetics, particularly the research of Vavilov, became world famous. The attack by the agronomist Lysenko on 'Mendel – Morgan – Weismann' genetics as idealist and reactionary, anti-Darwinist and anti-socialist, was not only based on inadequate scientific data, it was refuted by Marxist biologists both in the Soviet Union and the West. The success of Lysenko's attack, from the late 1930s onwards, when Vavilov was arrested and subsequently died in captivity, depended partly on Lysenko's association of the orthodox geneticists with the racialist views of the Nazis. Even more important was the appeal of Lysenko's claim that a socialist biology, based on the inheritance of acquired characteristics and later of the denial of competition between plants of the same species (for instance his insistence that trees of the same species co-operated with one another for mutual survival, à la Kropotkin, and should therefore be planted in close-packed clusters, where in reality they generally failed to survive), would transform the backward nature of Soviet agriculture. Lysenko claimed that this biology derived from the experience of the people – he was himself of peasant origin – and that it contrasted favourably with Vavilov's bourgeois approach which could not promise instant results.

So far as the debate in the Soviet Union was concerned, and despite the arrest of Vavilov at the end of the 1930s, the arguments were not really over until 1948, when following a full-scale conference of geneticists, agronomists and philosophers of science – in which the geneticists had much the better of the argument – Lysenko was able to reveal that his line had the support of Stalin himself. Virtually all his opponents were then forced to recant. The bureaucratic pressure towards narrowly conceived 'ideological purity' in science did not slacken in the Soviet Union until 1953.

The lysenko debates were published in full in Britain, and considerable argument took place amongst the Marxist scientists, reflected through several issues of the *Proceedings of the Engels Society*.[16] Several of the participants, certainly to their later embarrassment, swallowed scientific caution and accepted Lysenko whole despite their difficulties in coping with the manifest inadequacy at best and fraudulence at worst of Lysenko's experimental data. Few, other than Haldane – who accepted some and denied others of Lysenko's claims, and later quietly resigned from the Communist Party, after many years during which he had contributed a regular science column to the *Daily Worker* – seemed to have had any firm grasp of *both* the socialist *and* the scientific issues.* Haldane argued, predating Mao's observations on the correct way of resolving scientific questions, that Stalin's administrative interference was intolerable. Bernal, in contrast to Haldane, saw the debate in primarily power-political terms, arguing that to victimise the Soviet Union at the height of the Cold War was to help the capitalist system and that the Soviet Union was doing socialist science irrespective of the validity of Lysenko's work.

But the repercussions of the Lysenko affair could not be dampened down. One consequence of the embarrassment they caused was to speed the move towards the concept of 'neutrality' as a refuge for left scientists in Britain. If 'socialist biology' led into such open conflict with both intellectual integrity and scientific orthodoxy, then it was far better to abandon that particular way forward. Meanwhile the geneticists either left the Communist Party or relabelled themselves as 'radiation biologists'. Bureaucratic 'Diamat' (dialectical materialism) in science in the Soviet Union, and the industrialisation of science as an essential support of post-war capitalism in Britain and the United States, were together enough to inhibit further development along these lines for more than two decades.

Instead activity was diverted into other areas more directly related to the immediate political crisis: the shadow of the Bomb and the threats of the Cold War. Characteristically, different developments occurred in the United States and Britain. In the United States those most immediately affected by the Los Alamos/Manhattan Project experiences, the nuclear physicists, launched a dissenting journal of studies in disarmament and the social problems of scientists, the *Bulletin of the Atomic Scientists*, which was, because it concentrated on the nuts and bolts of disarmament rather than the political problems, to

---

*With the advantage of Levins' and Lewontin's work, it is possible to see that part of the problem for Marxists in assimilating and responding to the Lyskeno episode lay in the lack of a theory of cultural revolution. Hence they had no theoretical tools with which to develop an alternative to either a liberal eschewal of the whole affair or a 'stand up and be counted' solidarity with the Soviet Union. A critical solidarity was impossible.

prove readily capable of co-option. The American scientists had made their protest before the Bomb was dropped, and they failed. But their technological and scientific confidence was high. Strongly in the rationalist tradition of science, they seemed to believe that if only a technological fix could be found, disarmament could be achieved despite the underlying political and ideological issues which divide the United States and the Soviet Union. It is not surprising that the consequences of this approach were, on the one hand, the co-opted 'think tanks' and games researchers, the Kennedy liberals and even, eventually, the Department of Defense contracts, and, on the other, the elitist attempts to promote convergence with the equivalent Soviet elite by way of institutions such as the Pugwash conferences. Meanwhile unionisation and political activity amongst rank-and-file American scientists slumped.

In Britain some of the left scientists co-opted during the war, like Blackett, remained close to government thereafter. (It as hard to credit that Blackett, who ended in the 1964 Labour government's Ministry of Technology and as President of the Royal Society, had ever had his passport blocked to prevent him visiting the Soviet Union, for fear that he would be taking with him vital atomic secrets.) Those who remained outside government campaigned actively on the Bomb issue. The physicist E. H. S. Burhop, for instance, claims to have spoken at 150 meetings on atomic weapons within two years. Science was seen as a neutral bridge across the Cold War gulf. A Science for Peace group (along with peace groups for other cultural groups) was set up in Britain in 1950, but it had an active period of no more than a few years. At the same time, union work attracted a good deal of attention; the AScW passed through one of its more flourishing periods before becoming an almost exclusively technician's trade union at the end of the 1950s. It was scientists such as J. D. Bernal and W. A. Wooster who saw in the AScW the possibility of establishing an international scientific union. The World Federation of Scientific Workers was inaugurated in 1948 with a statement about the social role of science and a charter setting out the rights of scientific workers, seen as including some control of the utilisation and consequences of their science. The Communist physicist Joliot-Curie in Paris was the first president of WFSW, and its leadership has remained with the Anglo-French physicists ever since. Its major source of finance, because affiliation is on a *per capita* basis and the Soviet Union has the largest relevant union, has been Soviet.

The flurry of activity in the late 1940s soon died down however.

Science for Peace gently submerged and the early burst of activity over the nuclear weapons issue gave way to the quiescent despair from which CND and the New Left were to emerge so dramatically from 1956 onwards. Scientists were prominent in the Campaign – not merely the older ones but a new generation as well – but because of the essentially moral/pacifist, apolitical nature of CND, explaining the facts of fall-out did not demand a socialist commitment from the scientists; their role as representatives of a particular elite (along with clergymen, actors or MPs and their activity in pamphleteering or addressing meetings, did not necessarily either politicise or radicalise them. When they took on the government experts on radiation levels and hazards and won handsomely, they found themselves in their turn co-opted as experts on seismic detection of underground explosions, the monitoring of fall-out, radiation hazards, and so forth, travelling the same path as the physicists of the *Bulletin* had taken earlier in the United States. The rapid incorporation of Pugwash, following its inception in 1957, into an organisation fostering semi-official communication between the United States and the Soviet Union, demonstrated this process.

Disarmament was now a matter for experts, conservative or liberal-dissenting; the SALT talks would follow the Partial Test Ban Treaty separated only by learned articles in the specialist journals and almost unaffected by such irritating tremors on the geopolitical scene as Vietnam or Cuba. Co-option of the scientific elite in a part-time capacity was matched by the diversion of many rank-and-file peaceniks into research institutes ostensibly to work on issues of 'peace', 'conflict', and so on, often sponsored by the state, and leading to such subsequent ironies as conflict research on Northern Ireland in 1969 being sponsored by the Home Office. Thus, despite their goodwill and integrity, the growing self-awareness amongst scientists was safely contained.

CND therefore did not result in the politicisation of scientists as scientists; bomb-makers could be anathematised, for the *basic* scientific research on which the Bomb depended was largely complete; the escalation of *technology* did not find European university scientists – or many industrial scientists for that matter – actively involved in research which directly led to further Bomb production; the high technology and specialised scientific base of the Bomb ensured that the challenge that it presented to scientists was almost as satisfactorily moral as that it presented to the clergy.

It is perhaps more surprising that the emergence of the New Left at about the same time as CND was not to catalyse a much deeper analysis of the contemporary role of science, as it was to assist in the transformation of many other areas of cultural and political analysis in Britain. The *Universities and Left Review* leadership was uneasily conscious of the importance of science and the desirability of mobilising scientists; there were attempts to set up a New Left scientists' group, but the response was slight and the enterprise short-lived. Even on New Left platforms the scientists who spoke remained firmly in the social-democratic mould; any discussion which did occur concentrated almost exclusively on the use/abuse issue.

The ideological concerns which formed so important a theme even a decade previously remained buried; even on the left the mystification induced by Snow's high liberal concept of the 'two cultures' was to confuse debate for several years to come, for no socialist analysis of the cultural role of science was attempted by the New Left theorists. Science finds no discussion in *The Long Revolution*[17] for example, and Anderson's otherwise brilliant essay on *The Components of the National Culture*[18] excludes science from the analysis. The 1930s Marxist scientists are dismissed with a passing reference to 'the fantasies of Bernal', as leftist intellectuals who 'would be blown away by the first gust of the international gale'. This is precisely what did *not* in fact happen to the 1930s scientists, whatever was the case for the poets. Anderson's list of the components of British culture includes sociology, philosophy, political theory, history, economics, psychology, aesthetics, psychoanalysis, anthropology and literary criticism. His table of eminent *émigrés* who have contributed to British culture makes no reference to perhaps the most conspicuous group of all – the scientists, such as Born, Chain, Frisch, Gabor, Krebs, Peierls or Rotblat. Ironically, the political position of many of these figures would have sustained rather than refuted Anderson's analysis concerning the reactionary nature of those *émigrés* who came to Britain rather than the United States.

The point is that, a passive victim of the Snow thesis, Anderson cannot see them as 'cultural' figures any more than he can accept the serious contribution of the British Marxist scientists such as Haldane and Needham, who fail even to find a mention in his analysis. It is hard to escape the conclusion that there was a ducking of the question of science by the *New Left Review*.*

---

*See next page

In their concern to rediscover the young Marx, to pursue the themes of alienation and explore the French structuralist tradition[19], the New Left thus ignored the continuing role of science and technology in transforming, as well as being transformed by, the economic base. They showed, and still show, little recognition of the need felt and clearly expressed by Marx, Engels and Lenin, to understand the *content* of science and to assimiliate this content to their social analysis. It is not merely that the question of the relationship, for example, of DNA and the genetic code to Marxism was ignored by the New Left in a way which would have been inconceivable to Marx, but that very little recognition of the fact that the consequences of the application of science can now readily destroy the world appears in their writings. The long debates amongst Soviet philosophers as to whether the new science of cybernetics could be incorporated into the framework of dialectical materialism, or whether it required the addition of a fourth dialectical principle, which have occurred from the mid-1950s onward, find no echo among the New Left Marxists.

The consequences of this failure, together with the penchant of the *New Left Review* group in the late 1950s for collaboration with the left wing of the Fabians over a wide variety of practical issues, such as housing and poverty, thus left scientists without a sustaining alternative culture. Between this quiescence and the new phase of the radicalisation of science there was to be a discontinuity.

THE NEW SITUATION: INTERNATIONAL PERSPECTIVES

Vietnam was to be the symbol of this break. One of the earliest manifestations of the new wave was in 1966, when the International War Crimes Tribunal established by Bertrand Russell and Jean-Paul Sartre included scientists and doctors among its teams to collect evidence for its proceedings direct from Indochina. Their evidence on the experimental nature of the war and its use of new technologies, together with the growing heap of news reports and film filed by war correspondents,

---

*There has been a change since 1972 in the editorial board and policy of the journal. The natural world, whilst scarcely at the forefront of the *New Left Review's* interests, does now receive discussion. New Left Books has, for example, published Alfred Schmidt's *The Concept of Nature in Marx* (London: NLB, 1971) and Dominique LeCourt's *Marxism and Epistemology* (London: NLB, 1975). We are, ourselves, indebted to the *New Left Review* for the Enzensberger article which forms Chapter 9 of *The Political Economy of Science.*

compelled the scientists to speak out. Where in the past it has been the physicists, now it was the turn of the biologists, with the left and liberal scientists from a variety of disciplines coming to their support. For many of them the issues were moral rather than political; none the less some of the severest criticisms were made by scientists such as J. Mayer, Harvard public health expert who, in a carefully argued letter to *Science* was to describe the defoliation and crop-denial campaign as genocidic.[20] But while a few of the older scientists were co-opted (as when Mayer became one of Nixon's scientific advisers) others, especially the younger ones, moved rapidly from a position of liberal concern to a recognition that the social order itself needed radically changing. It was not long before they discovered that what Eisenhower had called the industrial – military complex had become the industrial – military – scientific complex and that the universities were deeply involved. This was a unique feature of the American situation, and attempts to trans-late it into other societies were to come to grief. American universities, ever since the old land-grant colleges, had been heavily dependent financially on, and closely controlled by, big corporations. Following the 1939 – 1945 war, a greater and greater proportion of spending on science and technology became carried by the Federal government (reaching some \$17 billion by the end of the 1960s); however, rather than adopt the strategy of establishing Federal laboratories and a scientific civil service, as in Europe, in the United States the policy was towards contract research in the universities and industry. By the mid-1960s, by far the greatest proportion of university science was being done on Federal contract, often for the Department of Defense, and the linkage was most powerful in the elite, high-science, ivy-league colleges. Napalm, for instance, was developed at Harvard and tested on its football pitches by the organic chemist, Fieser; its development is described in his neutral-sounding book *The Scientific Method*.[21]

From the mid-1960s on, in campus after campus students discovered a network of Department of Defense research contracts and institutes: the Institute for Cooperative Research (biological warfare research) at Philadelphia in 1965, the Stanford Research Institute a year later, the Lawrence Radiation laboratory and the existence of a whole class of 'classified' PhD work at MIT, and so on. By 1968 *Viet Report* had published lists implicating practically every university in the country, and a variety of 'workshop' groups began devoting themselves practi-cally full time to the collection of information about the involvement of universities and companies in war-related research. The work presented a natural rallying point for political struggle; only when, by 1970, most of the research had been driven off-campus did it become apparent to many that all that had happened was that the research was continuing under a different institutional hat — campus activism could win some victories, but the over-all effect was slight.

Nevertheless, the campaign had the effect of radicalising a proportion of science students and scientists; responses ranged from the liberal (the 'Union of Concerned Scientists') though to the more radical groups aligned with the left of SDS (Students for a Democratic Society). MIT held a brief research strike on 4 March 1969 which soon spread to other universities,[22] the biochemist who had worked with the Russell Tribunal, J. B. Neilands, ran a lecture course at Berkeley on the Social Responsibility of Scientists,[23] some scientists worked within their profession, by attempting to make, for instance, the American Physical Society discuss the use of physics in the Vietnam War and to pass appropriate resolutions. Other professional groupings became locked into the ABM controversy, which, because of open confrontation over the interpretation of evidence on the effectiveness of the ABM, was by 1972 to have split the U.S. Operational Research Association over the very issue of neutrality and objectivity in science which was the attack point for many on the left. Radical activity became more concerned with picketing scientific meetings, particularly the vast annual jamboree of the American Association for the Advancement of Science (AAAS), raising issues both of the uses of science and of the elitism of the platform scientific speakers. One of the most durable organisations to arise over this period has been SESPA (Scientists and Engineers for Social and Political Action), which, in a loosely federal structure, runs the gamut of the radical political spectrum from concerned liberals to movement revolutionaries. Its primary concerns have become a self-activating analysis of the role of science under capitalism in the United States and a continued concern with elitism in science (its racism, sexism, and so on). For this group, the urgent necessity is to use their science 'for the people', in the service of the blacks, the Puerto Ricans and other minorities, to turn the weapon of science against imperialism and capitalism.[24]

One development in the United States which has achieved considerable success in diverting the movement from its attempts to analyse the role of science has been the mushrooming growth of the pollution movement, suggested by many as having been fanned as a deliberately mystifying attempt to divert campus activists from the Vietnam War.

With the fragmentation of SDS, it indeed became the case that many who would have two years earlier been in the anti-war movement had turned their attention to the collecting of non-returnable bottles, campaigning for zero population growth or adopting the reactionary, anti-rationalist stance which we identified at the beginning of this chapter as one response to the breaking of the myth of scientific neutrality. The politicisation of pollution (as an inevitable product of capitalism) has scarcely checked this tendency.

In France, Italy and West Germany the politicisation of science and scientists developed slowly, and in the wake of the events of May 1968. The reasons for the slow development of the movement are probably to be found partly in the relatively backward state of science in the rest of Western Europe compared with Britain and the United States; only France was a nuclear power, and that only recently — and much less was being spent proportionally on defence research. Within the European cultural tradition 'natural science' is not separated off from the social sciences, or for that matter the rest of the intellectual superstructure — all are part of *Wissenschaft*. Where the cultural web is relatively seamless, to isolate a component in order to examine its uses and abuses is a less convincing exercise; instead the critique is more naturally generalised to relate to the social and economic base. On the other hand, the very different university structure of Europe, which, deriving from the Napoleonic reforms, places great authority and power in the hands of an undemocratic professoriat, reducing non-professorial researchers to a semi-technician status, has contributed to the development of militant teachers and scientists' trade unions, in contrast to the effete AUT and the weakness of the ATTI.

Thus when issues related to production have occurred, because of the stronger Marxist tradition in Western Europe, then they have been located firmly as part of working-class struggle, rather than as straightforward consequences of the use/abuse of science. In general the political activity of European scientists in the wake of the events of May 1968 concentrated on the nature of the work situation itself, with experiments in direct control of laboratories and the organisation of research.[25] University and CNRS (Centre National pour Récherche Scientifique) laboratories were occupied following May, but with the ebb of that high tide the sharply posed questions rapidly blurred down to questions of pay, status and the over-all reform of the French educational system — which have dominated the French university scene for almost the whole of the subsequent period.

Because of the continuing social and political crisis in Italy the consequence of the occupations there were more profound. The most conspicuous was the occupation of the International Genetics Laboratory at Naples in 1968. There the key questions were: Who controls the laboratory? Who owns the research? Can the elitist tradition of science, whereby the "great men" gain prestige as a result of the work of their research assistants, technicians and laboratory cleaners, be broken down, and, if so, could science continue to be done? Work stopped in the laboratories whilst the roles of technicians and cleaners − and the local community − in planning experiments and deciding on research directions, the relationship of experts to non-experts in society, was debated. Eventually the Director, Buzzati-Traverso, resigned (he went to work at UNESCO instead, where it would no doubt prove more tranquil) but the attempt to create socialism in one institute gradually petered out.[26] But the debates were to be repeated in many other places, and gradually the discussion of science and technology entered the working class movement.

THE NEW MOVEMENT IN BRITAIN

In Britain, the first ripples of the new wave began in 1967, in response to the Vietnam War. At first, much as in the earlier style of CND, scientists lectured and wrote pamphlets and articles exposing the nature of the war, making only little attempt to work within and organise their own occupational group. One early move to take this second step occurred when a small group of scientists, working in conjunction with the Angry Arts week at the Roundhouse in the summer of 1967, sponsored a meeting called 'Vietnam: the Abuse of Science'. This was followed up by a continuing group who set up an international meeting on Chemical and Biological Warfare in February 1968. In general the style of the meeting was sedate and highly professional. However, some light was thrown on the nature of the 'neutrality' of much of contemporary professional science when a French doctor (M. F. Kahn) presented data from the NLF and from the DRV concerning the toxicity of the chemical weapon CS gas. The point he was making was the illegality of the US government's use of internationally banned poisonous gas. Some of the American scientists present were quite incapable of accepting this data; Hanoi or Vietcong science, by definition, was propaganda. This exchange probably did more to politicise the meeting than any formal analysis of the nature of American imperialism. In

1968 the public breach of the myth of the neutrality of science was beginning to be a means of raising the consciousness of the profession.[27]

Out of this meeting came a continuing and more radical group which saw its task first as the on-going development of the anti-CBW campaign, and then as the drawing together of their experience into a wider movement, the British Society for Social Responsibility in Science (BSSRS). A child of its time, BSSRS (and it was soon to dislike being called by its full name) saw its initial tasks as self-education for scientists concerning the control of science and exposing the abuses of science, with the goal of politicising and mobilising increasing numbers of scientists and science students.

The society was at first mainly active on the campus, yet science students themselves were often less involved in its activities than those from other disciplines, almost taking a pride in their alienation, explaining themselves by reiterating 'we are only small parts in a machine'. Virtually only the religiously committed were uncrushed, because they saw both the problems and the solutions in purely moral terms, and therefore demanded only a simple, voluntaristic answer, a personal moral renunciation of doing particular sorts of science. The rhetoric of the 'scientific worker', which had been so energetically made in the 1930s and 1940s by the Marxist scientists (themselves typically of elite origin and with all the confidence of the scientific elite) became a reality to the students of the 1960s and 1970s. The industrialisation of science had helped produce a proletarianisation of scientists, resulting in a deepening gulf between the ostensibly non-alienated 'responsibility' of its elite and the alienated apathy of the masses.

Always more of a movement than a membership organisation, BSSRS early on adopted a fairly standard centralised constitution which it then proceeded to ignore, maintaining a federal reality of autonomous local groups. To some extent the multiplicity of political stances* at once the strength and the weakness of BSSRS, was to do with the nature of the issues with which it was concerned. The expertise required to contribute to the CBW debate, as the chief preoccupation of BSSRS in its early days, meant that scientists with a very wide range of political beliefs were able to work together in technical discussions.

---

*The first National Committee included CP members and fellow travellers, Labour Party members and supporters, liberals, Trotskyists, counter-culture anarchists, and non-aligned socialists (with a certain amount of movement between categories, some ending up as Maoists).

Because of this need for expertise, the Society began by accepting mentally much of the elite structure of science. It acquired a distinguished sponsorship of left and liberal scientists and held its inaugural meeting in the Royal Society.[28] Yet the libertarian socialist tendencies, which were later to manifest themselves much more strongly, both with BSSRS and other parts of the scientists' movement, were given some expression by the society's willingness to encourage non-scientists to join. Outside the scientific community (and in 1976) this modest step as an indicator of a socialist tendency may seem absurd, but the rigidity and hierarchical nature of the laboratory is perhaps not fully realised beyond its confines. To cite an equally absurd but illuminating example, a profile published by a scientific weekly of a left-wing professor described his willingness to let his students call him by his first name as an index of his revolutionary stance. A combination of an inability to confront the elitism of science, and the need felt to finance itself and expand by playing the media rather than building a movement from the grass roots, characterised one continuous aspect of BSSRS's early history.

From the beginning BSSRS had related its science to revolutionary movements for national liberation. In response to an appeal from Derry Civil Rights workers after the RUC attack on the Bogside in August 1969 it sent a team of natural and social scientists to Derry to collect evidence on the effects of CS.[29] But their preoccupation with getting the science right and hammering the government experts meant that, to a large extent, they became caught by the same trap of neo-professionalism which had caught the left physicists against the Bomb. On the issue of CS, BSSRS made the bad mistake of merely playing the media and submitting its evidence to the government's Himsworth Committee. It failed to write the necessary agitational and educational pamphlets.[30] It was left to others working directly with the Irish movement to do this.* Not surprisingly, BSSRS's National Committee, weakened by addressing the wrong audience, lost itself in the liberal 'gas or guns' debate, and was doomed to do little more than express moral concern about an unpleasant technology which was in actuality an inherently imperialist weapon.

This tendency to play the elite/media game continued to represent one aspect of the Society's schizoid personality. Even although it eventually did declare against both CS and rubber bullets, it still, even in

---

*This changed with the publication of the joint BSSRS and Troops Out Movement pamphlet *The New Technology of Repression* (1974).

1972, felt it must give a commentary on the Rothschild report, which contained proposals to confirm the industrialisation of science by further limiting the autonomy of the Research Councils, building an openly declared 'customer-contactor' relationship into the funding of large areas of hitherto 'pure' science. The Society's comment betrayed the confusion still central in its structure; it merely opposed the report for wanting to spoil science by placing it, even more crudely, at the disposal of capitalism, rather than accepting head-on the ideological challenge of confronting capitalism with the demand that science should instead serve the interests of the working class.

The other aspect of the Society was represented by the activities of its more rank-and-file membership. After the inaugural meeting, in April 1969, groups and local societies began to emerge on many campuses and in a few towns and by the first AGM, in November, membership had grown to nearly 1000. Predominantly the membership was academic, but the variation in style between the differing local organisations was immense.

Some were early captured by a scientific establishment that wished to do little more than express liberal concern. The Edinburgh society, for example, was initially sponsored by left scientists, but was seen by the university partly as a welcome evidence of social concern and partly as a potentially radical rival to the existing Science Studies Unit; it was soon safely captured by the Unit. The alliance which easily developed between the liberal and the old left scientific establishment emerged in sufficient strength to exclude any radical intervention, and Edinburgh SSRS was soon to settle down as a pleasant lunch-club activity, unlikely to raise any serious critical issues.*[31] The Cambridge group,[32] on the other hand, never really received other than very nominal support from its very substantial scientific elite, and was freer to work with school children, and to begin work on a book on 'race and intelligence'.† Others such as the group at University College (London) sought to study sponsored research on the university campus, but the line which had been so fruitful in the United States was much less so in Britain, where the very large amount of defence research that is done is primarily located in government research establishments rather than university laboratories. Indeed, even where the defence research did exist, as in

---

*Eventually a rival 'Science for the People' group established itself around the issue of unionisation of research students.

†This work was eventually carried though by the London-based Campaign on Racism, IQ and the Class Society', and the Cambridge group became dispersed.

Porton contracts held by Southampton University, the student movement was slow to understand their significance, ignoring the contracts
when their details were published by BSSRS and on the front page
of the Sunday Times, which were only seen as a political issue when
social, rather than natural, science students liberated them from an
office during an occupation.

This reflected the very limited number of science students who were
directly involved in such groups as the Revolutionary Socialist Students'
Federation. By the winter of 1969 a sufficient group had emerged to
convene two meetings of a Federation of Revolutionary Socialist
Scientists (FORSS) at Manchester and then Warwick. But although
the group managed to publish six issues of *Red Scientist*, a small,
primarily theoretical journal, before its demise, and although individual
members of FORSS worked within BSSRS, the organisation was
left to too few people, primarily based in Manchester, and soon
disappeared.

BSSRS's attempts to move from the use/abuse issue to the broader
one of the neutrality and role of science under capitalism were
characterised by two large meetings, one in Cambridge in summer 1970,
the second, five months later, in London. The theme of the Cambridge
meeting was Race, Class and Intelligence, the immediate purpose being
to air and refute Jensen's claims concerning the genetic inferiority of
the IQ of US Blacks.[33] The consequence was to develop the first
extended attack on bourgeois ideology in science (the mass media
widely reported this attack as an attempt to 'suppress the search for
truth' in science).[34] Even though the first attack was in social psychology, where ideological forms are more plainly apparent, the next was
to tackle a slightly 'harder' science. This was in the winter of 1970, at
the meeting on the Social Impact of Modern Biology, where, because of
the presence of the Society's President, the Nobel Laureate M. H. F.
Wilkins, on the organising committee, it was possible to command the
participation of most of the luminaries of molecular biology. Although
almost certainly the huge audience of 800 came and stayed for the
three-day meeting because of the presence of these stars, the discussion
increasingly criticised the organisers for their elitism. The American
molecular biologist Jon Beckwith's paper, echoing this feeling from the
radical wing of SESPA, was received with enthusiasm and the papers
raising issues of ideology in biology reopened the discussion which had
been closed since the Lysenko affair.[35] Within the Society relations
between those who wanted to explore the possibility of a socialist

science and those who preferred to concentrate on issues of use and abuse became strained. The alliance of left tendencies and liberals which the lack of political clarity in the scientists' movement had hitherto permitted began to break down. Was BSSRS a broad front, or should it (as we ourselves had argued) become clearly committed politically? The federal structure prevented the issue being totally disruptive, each local group could do its own thing, but on the National Committee, still lured by media politics, the issue was unresolved.

Anti-elitism as a strand within BSSRS's politics broke through the summer of 1970, when, modelled closely on the SESPA demonstration at the Christmas 1969 meeting of the American Association for the Advancement of Science, a group within BSSRS attended the meeting of the virtually moribund British Association. Street theatre and some interventions at lectures brought relief from the boredom of the BA meeting which, while of little significance to scientists, continued to command an inordinate amount of space in newspaper coverage. The journalists, with an almost audible sigh of relief, joined the BSSRS criticisms to some of their own. Few scientists attend BA meetings except to give papers, so there was little recruitment of scientists to the Society, but the issue of the neutrality of science received its first public hearing at a six-hour 'teach-in'. In some ways the impact of the meeting was blurred by the proposal from the libertarian socialist Solidarity Group to institute a 'Hippocratic oath' for scientists.[36] The 'Durham Resolution' which might have been a good consciousness-raising move a few months earlier, seemed, in the summer of 1970, to quote the liberal physicist, John Ziman, so woolly that any defence scientist could sign it in good faith. Perhaps the irrelevance of the BA meeting to science was plain; for, unlike SESPA, BSSRS has not so far repeated the exercise.*

By 1971 the most important issues were seen as those of ideology in science and self-management. A self-management conference was held in 1972 drawing participants from all over Britain,[37] but, while the French experience was discussed, the essential difference was that in France the debate had taken place in the laboratories themselves as part of the context of the May events.

---

*Activists within BSSRS have disagreed as to the significance of the 'Durham Resolution'. Jonathan Rosenhead regarded it as a watershed separating the new radical BSSRS from its liberal precursor (*New Scientist*, 54 (1972) p. 134); Alan Dalton sees it as indicating that 'the early philosophy and action of BSSRS related mainly to moral issues', *Chemistry in Britain*, 11 (1975) p. 107.

As BSSRS began to discuss the concept of 'community science', whereby science was to be involved at the neighbourhood level in the struggle against pollution and other health hazards, other similar groups scattered around the world began to engage in locally based activities. Just as BSSRS engaged in the campaign against the 'Battersea Smell', so scientists were mobilised within community struggles in Southern France and Bremen against the danger of proposed nuclear power stations, disposal of sewage at Barcelona, and with the fishermen against mercury poisoning at Minamata in Japan. For the British, the initial danger lay in the word 'community', which bears in its train all the problems of localism and the verbal dissolution of class antagonism. Science for the 'community' was acceptable in a way that science for working-class and liberation movements was not. Community science, like community medicine or community care, was thus particularly welcome to sophisticated ruling-class interests. The problem remained of distinguishing between voluntaristic, individualistic conceptions of the role of science and the scientist under capitalism, and those which recognise its structural situation.

PERSPECTIVES FOR THE RADICAL SCIENCE MOVEMENT

In writing the first version of this chapter we argued that it was necessary to analyse the social function of science, and that, at least in Britain, Marxists had failed to develop this analysis. The scientists have had to stumble, self-taught, through the use/abuse issues until they arrive at the core problems of ideology, alienation and elitism.

Programmatically, we argued that the items which must now be on any agenda for action include:

(1) The systematic exploration of the ideological components of capitalist science. As since the Lysenko period this field has been to all intents and purposes abandoned by orthodox Marxists, the task of recreating this understanding is immense. It is necessary to show both the social determinants of science under capitalism which direct research in the interests of profit and imperialism, and the mechanism whereby these determinants are internalised by the researcher, determining paradigms, experiments and interpretations. And it remains necessary to grapple with at least trying to ask (for we cannot answer) the question: what would a socialist science be like?

(2) The attempt, within the existing framework, to create a contra-science. The phrase 'science for the people' presently carries more emotional charge than substance. Among its implications are the arming of the working class with knowledge, the scientific information with which to demystify and defeat the assumption and utilisation of expertise by the ruling class; and the use by the scientific workers of their skills as an integral part of the struggles of working-class, Black, women's and community groups in projects ranging from the development of insurgency technologies to the monitoring of levels of possible pollutants. But in exploring these possibilities, we must always be aware that the science and technology scientists are putting at the service of the people derive from — and are likely to remain — bourgeois culture. In the struggle to make science for the people, science itself will be transformed.

(3) This transformation carries with it the breaking down of the barrier between expert and non-expert; socialist forms of work within the laboratory, making a genuine community in place of the existing degraded myth, must be matched by an opening of the laboratories. The Chinese attempts to obliterate the distinction of expertise, to make everyone their own scientist, must remain the aim; laboratories which are integrated into their local communities, and within which the benches and equipment become open for all to develop their projects, would help break its stubborn elitism, and transform the doing of science.

All these developments must be set into a political framework, which recognises both their long-term strategic significance as experiments in what socialist forms of science will be like, and their more immediate role in the raising of consciousness amongst scientists. To fail to appreciate this dual role would obviate their significance.

It is precisely because of the need for this political framework that the innovative developments of recent years need consolidation. The task can no longer be left to organisations of scientists, operating in a liberal/libertarian mould, or small theoretical groups. Scientists now need to bring their science into the area of activity of the Marxist groups, and the groups to accept their responsibilities for political work in science. This collective development is the way forward to a much wider and deeper radicalism amongst scientists than ever before.

POSTSCRIPT

The four years since 'The Radicalisation' was first published in *The Socialist Register* have been a period of considerable change within the radical science movement as it has confronted the deepening economic crisis which has affected in varying degrees the entire capitalist world economy.

While unemployment has been sharpest for shop-floor workers, the scientific and technical workers have not been immune. Science-based industry, the darling of the economic-growth theorists of the 1960s, has suffered — though not as sharply as the traditional sectors. But economic crisis and unemployment of themselves do not produce radicalism; they may only increase economistic trade unionism with its emphasis solely on the defence of the wage packet — even at the expense of the right to work. In Britain, as elsewhere, the scientific unions have all too often agreed redundancies, and while protecting their own members neglected the rest of the working class. One outstanding exception has been the campaign of the shop stewards of the Combine Committee of Lucas Aerospace around their own 'corporate plan', designed to protect jobs and redirect the skills of Lucas workers towards socially useful technologies. In general such developments have come from the workers in struggle themselves and the radical science movement, while enthusiastically welcoming such initiatives, has as a result of its academic and student composition not been directly involved. Everywhere the ebb of the high tide of the student movement has left activists aware of a new and more complex situation; there has been a conscious search for theory, an attempt to interpret the slogans of science for the people in terms of an understanding of the historic role of science under capitalism, and, in many countries, a turning away from the spontaneism of the 1960s towards the need and the difficulties of alliances with the trade-union movement and community groups. Organising around questions of employment, industrial health and pollution, have become key areas.

Action for the movement has ranged from the essentially defensive through to the discovery of new spaces for political activity. At the defensive end of the spectrum have been the campaigns in West Germany for the protection of individuals from political persecution and loss of jobs in a situation in which legal proscriptions have been invoked against communists, members of revolutionary organisations and even left/liberals having jobs anywhere within the educational system.

The internal development of the movement has also presented its own problems. For example, within the strongest and most energetic of the radical science groups, *Science for the People* (SftP) in the United States, there has been a struggle involving some of the sectarian organisations which have found in SftP an arena of political work which substitutes for the inability of revolutionaries in the United States to build a working-class base. Whilst the forces which have tended to isolate revolutionary parties from the workers in the United States derive from the particular experience of the US working class, especially since the Cold War period, the consequence for SftP has been the attempt to inject a single-minded ideological purism into a social movement whose goals are necessarily more diffuse.

At the theoretical level the struggle has focused around the analysis of the type of work open to an organisation like SftP, expressed first as the choice between consciousness-raising and party-building, and second as the identification of the main enemy in the context of the struggle within the United States. One of the contending lines, reflecting the thinking of the October League, has been to identify imperialism not merely as *a* main enemy but to all intents and purposes as *the only* enemy for struggles both within and outside the United States. Whilst this has led to a major involvement of SftP in work around Black and Third World peoples' struggles, it has also resulted in a playing down of activity on general working-class issues like industrial hazards and technological unemployment. While SftP is in no danger of returning to the ill-concealed paternalism of demands for unity between Black and White which in fact always meant the subordination of Black interests to White ones, it does move towards a new danger of laying such stress on Black and anti-imperialist struggles that it excludes the possibility of an alliance between all anti-capitalist and anti-imperialist forces.

practical levels, varying from area to area: confrontation, agitation, pamphlets and 'learned articles' against the racist science associated tion, pamphlets and 'learned articles' against the racist science associated with the names of Jensen, Herrnstein, Shockley and, most recently, E. O. Wilson, author of an influential ethological tract entitled *Sociobiology* (see Chapters 6 and 7 of *The Political Economy of Science*). Off campus, there have been programmes of school teaching material and direct involvement in the bitter campaign around bussing in the Boston area.

Specific issues on which SftP has been able to intervene and score notable successes have remained primarily in the area of ideological struggle around elitist and sexist practices within the science knowledge industry itself. Thus campaigns for the restriction of the uses of psycho-surgery (which, as its use extends to Europe, the movement elsewhere will have to pick up too) and the termination of the XYZ screening programme in Boston, in which male babies were given a chromosomal test and, depending on its results, parents informed that their sons might become 'psychopathic criminals', were both successful. These issues involve both ideology in science and its role as a generator of social-control technology and the genetics of oppression and are discussed in more detail in *The Political Economy of Science*, and they have repre-sented areas in which it has been particularly easy for SftP to work because of the nature of its membership. Similarly, it had been possible to become very directly involved in the struggle against sexism. Women's groups have developed programmes for self-care in medicine, and this work in health has extended to a generalised attack on the capitalist and sexist nature of US medical practice. Particular centres, such as the Boston Women's Center and Health Pac in New York have acted as focal points of information gathering and dissemination in this process, linking the women's movement with health and science. The strengths of the women's movement in the United States has exposed male activists within SftP to a critique which has helped considerably to shift its own internal practice, and far more enduring links with the women's movement have been built than has been possible with, for example, the overwhelmingly non-campus Black and Third World movements.

It is not just in the United States that links with organised labour, even in the science-based industries, have been hard to achieve. None the less, where they have been built, as in Rome and Palermo, London and Birmingham, they are now concrete, involving joint activity in, for example, struggles to reduce health hazards at work. Although the gains are modest, they mark a significant departure from the politics of tokenism; of the trade-union official speaking on the left scientsts' plat-form and vice versa. In addition, particularly in France and Italy, the movement has been successful in developing the contradictions within the scientific knowledge industry itself. During and immediately after 1968 the emphasis was on exposing hierarchy, the oppressive ideology of science, the irrelevance of bourgeois science for the lives and hopes of working people; an exposé which was linked to the development, through a collective practice, of an alternative science and technology. The idea of the fully fledged collective as an experiment in utopian socialism soon withered, but work focused around the lab situation has remained for some strands of the movement in the United States and in Europe. In Italy there was a brief flowering of 'Science for Vietnam' collectives in many university laboratories, but in general by the mid-1970s the collective had become less important as an end in itself – an alternative – and more important as an innovatory oppositional base from which to combat existing scientific practice and knowledge. Collective practice has become less the mark of the essentially anarchist counter-culture and instead belongs more to the socialist contra-culture.

This collectivist practice has been the core of the on-going work of the movement in France. Here the themes first raised in 1968, of hierarchy, elitism and alienation in laboratory practice, have continued at the forefront of activity. They are the dominant concerns of the collective which produces the magazine *Impascience*, largely but not entirely Paris-based, which has attempted to weld together the exposure of scientistic ideology with the very direct experience of oppressive practice as it affects technicians, secretaries, auxilliary workers and – of course – women, who are cast in all these roles as the under-labourers of science.

At the time when we first wrote 'The Radicalisation of Science', the movement in Britain, with BSSRS a prominent but not the only strand, seemed to have swung away almost entirely from work amongst scientists themselves, or even campaigning on specific issues (apart from on-going concern with CBW and riot-control technology), towards a rather loosely defined 'community politics', alternative technology and the development of self-activating, self-sufficient rural or urban communes. The anarchistic burgeoning within BSSRS was felt to be sufficiently anti-establishment that layers of liberal elite scientists peeled themselves away, setting up the entirely incorporated'Council for Science and Society', from which they could preach unhindered the doctrine of the neutrality of science.

Over the period till the winter of 1973 the connections between the radical science movement as expressed by the mainstream BSSRS and its magazine *Science for People* on the one hand, and working-class struggle on the other, became increasingly attenuated. The much smaller Marxist element sought to work with the Troops Out Movement and collectively wrote a pamphlet on *The New Technology of Repression* in Northern Ireland (later to be a book). Initially the language of class sounded heavy to the ears of the counter-culture, but by the end of the winter the miners' strike had not only brought down the Conservative government but had made manifest the reality of class relations and class politics.

Thus the economic crisis in Britain and the changing experience of the movement has led steadily back towards a more directly political involvement in which both libertarian and Marxist strands have participated, and have moved in the process closer together. For the former the concept of 'organisation' is now viewed less hostilely, whilst for the latter there is distinctly less certainty about the theory of the vanguard party. Constructed most optimistically, there can be seen in this coming together the beginnings of a non-vanguard revolutionary practice.

The new activity has centred partly around theory – for instance, the emergence of a collective producing *Radical Science Journal* and the development of on-going study groups – and partly around practice. The links with wider political and class struggle were strengthened first by BSSRS endorsing the *New Technology of Repression*, and second when it was recognised that an effective space for action which united general concerns over pollutant technology with issues which directly affected working-class people lay in the area of industrial health and work hazards, this tendency is shown within BSSRS both in links with shop stewards' committees around specific hazards in particular factories and the publication of a series of pamphlets aimed at helping workers mobilise around hazards such as noise or oil mists.

The campaign around scientific racism, which was sponsored by BSSRS and *Humpty Dumpty*, the radical psychology magazine, was able briefly to achieve what had not been done in the United States – to come out of the campuses into shared work with Black groups against the labelling and schooling of children as educationally subnormal ('ESN'), and with the trade-union movement against racist practices which keep Black and Asian workers in the least skilled and worst-paid jobs. This period was shortlived as each left political group and partly developed its own 'unifying' trade-union organisation. The Campaign against Racism, IQ and the Class Society was thus forced to limit its work to combatting the ideas of class and race supremacy in the schools.

Nor has the insistence of orthodox Marxists as seeing all struggles as subordinate to those of the (overwhelmingly male and, by implication, white) industrial proletariat helped the movement in Britain to overcome its own male chauvinist practices. Achieving a less sexist practice thus not only has immediate implications for how women and men work together politically but also for the theory the movement is trying to create. One factor which is likely to accelerate this process is the pressure from feminist groups such as Women and Science (responsible for one of the two women's issues of *Science for People*), which sees itself as primarily part of the women's rather than the science movement. Aware of the great isolation of women in science, and of almost all women from science, they see their tasks as demystifying science for women (including themselves) and exposing the male dominance of science in both its organisational and ideological forms. In the context of the economic crisis, where women are threatened with a return to their position within the reserve army of labour, the ideological role of the psychologism and biologism of people like Buffery, Hutt and Grey in 'justifying' woman's role as an inferior domestic being is seen as a particularly important enemy. The links with the women's movement, with its associated attack on the patriarchy, and on bourgeois ideology in general, will have the effect of advancing the radical science movement's own cultural critique and the analysis of the social-control functions of science which have in the last few years lagged behind its work in the area of science for production and profit.

It is perhaps in Italy that the highest level of integration of the radical science movement with general political struggle has been achieved. The spaces for action that have been found — work on industrial health in association with women factory workers and with agricultural workers in the Rome region, joint political/economic/scientific education programmes with members of the metal-mechanics union following the winning by the unions of an entitlement to '150 hours' of study-time for the workers, and so forth — have not been too different from those that have developed in Britain. But, as with the struggle for the city in Italy, they have been much more theoretically informed. This both reflects the much higher intensity of the social and political crisis in Italy than elsewhere in advanced capitalist countries and that the theoretical critique developed by Manifesto and the practical struggles in which Lotta Continua have engaged (two of the principal revolutionary groups which have been associated with the radical science movement) constitute, even in their organisational separation, a theoretical and practical whole.

It goes without saying that the radical science movement takes different forms, achieves different levels of theoretical and practical development in different countries. What is also true and important about the movement is the way in which, despite all its unevenness and hesitancies, it increasingly occupies itself with both the hard questions of the political economy of science; science for accumulation and profit, and science for social control; and at the same time seeks to develop a new personal practice in which the social relations of the movement anticipate and rehearse the social relations of a new and just society.

2

The Problem of Lysenkoism

Richard Lewontin and Richard Levins

The Lysenkoist movement, which agitated Soviet biology and agriculture for more than twenty years, and which remains attractive to segments of the left outside the Soviet Union today, was a phenomenon of vastly greater complexity than has been ordinarily perceived. Lysenkoism cannot be understood simply as the result of the machinations of an opportunist-careerist operating in an authoritarian and capricious political system, a view held not only by Western commentators but by liberal reformers within the Soviet Union. It was not just an 'affair', nor the 'rise and fall' of a single individual's influence, as might be supposed from the titles of the books by Joravsky[1] and Medvedev.[2] Nor, on the other hand, can the Lysenko movement be regarded, as some ultra-left Maoists do, as a triumph of the application of dialectical method to a scientific problem, an intellectual triumph that is being suppressed by the bourgeois West and Soviet revisionism. None of these views corresponds to a valid theory of historical causation. None recognises that Lysenkoism, like all non-trivial historical phenomena, results from a conjunction of ideological, material and political circumstances, and at the same time is the cause of important changes in those circumstances. It is not particularly surprising that bourgeois commentators have such a view of the Lysenkoist movement, for it is entirely within their tradition that major historical changes can be the result of individual decision and caprice of powerful persons or of unique historical accidents with no special causal relations. Thus, Joravsky, whose book calls attention to a great many of the complex forces that contributed to the Lysenkoist movement, nevertheless explains the rise of Lysenkoism as essentially the consequence of 'bossism' in which the political 'bosses' of Soviet agriculture. including the 'supreme boss', Stalin. embraced an incorrect scientific doctrine in a blind and capricious flailing about in search of solutions to Soviet agricultural problems,

problems which they themselves had created by their irrational programme of collectivisation. It is rather more surprising that socialist
writers, who are supposed to know better, are equally narrow in their
understanding. If they are liberal reformers like Medvedev, they view
Lysenkoism as a boil on the body politic, a manifestation of the Stalinist
infection that is poisoning a potentially healthy revolutionary organism.
If they are like some Maoists, they restrict their view to the philosophical aspects of the problem using Mao's essay *On Contradiction* in an
attempt to prove, as Prezent claimed,[3] that Mendelian genetics is
incompatible with the principles of dialectical materialism, and that a
rigorous application of dialectical method will lead to Lamarckist
conclusions.[4] We must reject both of these viewpoints as too narrow.
Of course it is true that authoritarian political structures in the Soviet
Union and bureaucratisation of the Communist Party had a powerful
effect on the history of the Lysenkoist movement. Of course it is the
case that the methods and conclusions of science contain deep ideological commitments that need to be re-examined. But there are other
factors as well that were part of the material and social conditions of
the Soviet Union and which were integral in the Lysenkoist movement.

The Lysenkoist movement of the 1930s–60s in the Soviet Union
was an attempt at a scientific revolution. It developed in the context of
the pressing needs of Soviet agriculture which made the society recep-
tive to radical proposals; the survival of both Lamarckian and non-
academic horticultural traditions on which to draw for intellectual
content: a social context of high literacy and popularisation of science
which made the genetics debate a public debate; an incipient cultural
revolution which pitted exuberant communist youth against an elitist
academy; and a belief in the relevance of philosophical and political
issues which made the discussion take place in the broadest terms. But
it also took place in the context of the encirclement of the Soviet
Union, the Second World War and the Cold War. Administrative repressiveness and philosophical dogmatism increased; opportunists jumped
on bandwagons, the cultural revolution aborted.

In the end, the Lysenkoist revolution was a failure. It did not result
in a radical breakthrough in agricultural productivity. Far from over-
throwing traditional genetics and creating a new science, it cut short the
pioneering work of Soviet genetics and set it back a generation. Its own
contribution to contemporary biology was negligible. It failed to
establish the case for the necessity of dialectical materialism in natural
science. While in the West it was interpreted merely as another example

of the self-defeating blindness of communism, in the Soviet Union and Eastern Europe it is still a fresh and painful memory. For Soviet liberals, it is a classic warning of the dangers of bureaucratic and ideological distortions of science, part of their case for an apolitical technocracy.

Our own interest in re-examining the Lysenkoist movement is severalfold:

(1) The interpretation of scientific movements in terms of their social, political and material context, rather than the idiosyncratic, is a major task of intellectual history. More than other fields of historical research, the history of science is steeped in notions of accident and personal achievement as the motivating forces of history. The development of a materialist history of science is still in the future despite the pioneering work of Hessen and Bernal.[5] The Lysenkoist movement is recent and well documented, yet the major scientific issues between Lysenkoists and geneticists have been resolved by developments in genetics. Therefore the problem has the advantage of being both contemporary and yet of belonging to the past.

(2) There are important issues of the general methodology of science, and the relationship of scientific method to the requirements of practical application, that were raised by the Lysenko controversy and that remain open. We have in mind particularly the requirement of a control for experiments and the standard techniques of statistical analysis, both of which were challenged by the Lysenkoists.

(3) As working scientists in the field of evolutionary genetics and ecology, we have been attempting with some success to guide our own research by a conscious application of Marxist philosophy. We therefore cannot accept the view that philosophy must (or can) be excluded from science, and deplore the anti-ideological technocratic ideology of Soviet liberals. At the same time, we cannot dismiss the obviously pernicious use of philosophy by Lysenko and his supporters simply as an aberration, a misapplication, or a distortion dating from the era that is often brushed aside as 'the cult of the personality' (with or without naming the personage in question). Nor is it sufficient to note that Marxism has had signal successes despite Lysenko, with its pioneering work in the origin of life among its achievements. Unless Marxism examines its failures, they will be repeated.

In its last years, Lysenkoism was a caricature of the 'two camps' view of the world, in which the confrontation of bourgeois and socialist science was seen as parallel to the confrontation of imperialism and the anti-imperialist, socialist forces. Its absurdities could easily lead to a denial that there are two camps, a viewpoint which stresses the common ground of all science in a neutral, technical rationality independent of its uses. The present reduction of armed conflict would seem likely to strengthen this view of science at a time when we believe that the conflict within science must be made sharper as well as more complex.

Therefore this review is at the same time part of our own process of self-clarification.

THE PHILOSOPHICAL AND SCIENTIFIC CLAIMS OF LYSENKOISM

The main thrust of Lysenkoist research was the directed transformation of varieties (interpreted as the directed transformation of heredity) by means of environmental manipulation and grafting. This work directly contradicted Mendelian genetics. A second line of work emphasised physiological processes, which, while not formally incompatible with Mendelian genetics, were certainly alien to its spirit and were ignored by geneticists. This included such studies as the reversal of dominance, selective fertilisation, the influence of pollen on the flavour of fruit, which is genetically maternal tissue, and the physiological interactions in sperm and pollen mixtures.[6]

The main theoretical structure of Lysenkoism is the following.

(1) Heredity is a physiological process, a result of the whole lifetime of interaction between organism and environment.

(2) The assimilation of environmental conditions by the organism takes place in accordance with its own heredity. Suitable aspects of environment are selected and transformed; unsuitable aspects are excluded. In the course of development, the heredity programme unwinds like a spring, at the same time winding the spring for the next generation.

(3) If the environment is suitable for the normal expression of the heredity of an organism, that same heredity is reproduced in the reproductive cells. But if the environment does not permit the normal expression of the heredity, it will also alter the processes producing the heredity of the next generation.

(4) Factors which destabilise the heredity and permit its modification are: (a) altered physical environment as in vernalisation (a treatment intended to make earlier sowing and germination possible — see note 11);

(b) grafting, especially at very early stages of development, and with the removal of leaves making the graft dependent on its graft partner. Grafting is most effective if done early in development. Thus the transplanting of plant embryos to the stored seed nutrient endosperm of other varieties, or the production of genetically different endosperm by using mixtures of pollen, provides the most favourable conditions for vegetative hybridisation. The equivalent process in animals is the use of mixed sperm. Sperm which penetrate the chicken egg without actually fertilising the nucleus metabolise for a while and serve as an internal mentor or guide to development; and (c) hybridisation.

(5) The assimilation of nutrients and of external environment is dominated by the heredity pattern of the organism. But in sexual reproduction, each gamete is the environment of the other. Thus fertilisation is the mutual assimilation of different heredities. The result is especially labile and subject to environmental influence.

(6) The same cause that results in the production of an altered heredity or of new varieties — the exposure to a pattern of environment that cannot be assimilated in accordance with the old heredity — is also responsible for the origin of new species. Thus speciation is not a population phenomenon but an expression of individual developmental physiology. This is in keeping with the older Lamarckian view.

By and large, the philosophers sided with Lysenko, whose general approach seemed more plausible from the viewpoint of their interpreta-tion of dialectical materialism. The major philosophical issue was the Lysenkoist claim that the gene theory was metaphysical and the gene a mystical entity. From the earliest days of Mendelian genetics, the major books would make such statements as:

Germplasm, the continuously living substance of an organism. It is capable of reproducing both itself and the somatoplasm, or body tissue, in giving rise to new individuals. It is the Substance, or Essence, of Life which is neither formed afresh, generation after generation, nor created nor developed when sexual maturity is reached, but is present all the time as the potentiality of the indivi-dual before birth and after death, as well as during that period we term 'life' between these two events. The somatoplasm, on the other hand, has no such power. It can produce only its kind, the epheme-ral, the perishable body or husk, which sooner or later completes its life cycle, dies and disintegrates. The germplasm, barring accident, is in a sense immortal.[7]

Such statements, which geneticists brushed off as extreme views, were regarded by the Lysenkoists as extreme only in frankness and clarity but in no way contradicting the mood of modern genetics of the 1930s.

Geneticists would respond that textbooks do not reflect the real thinking of the working geneticist, that they obviously recognised the material nature of the gene, that otherwise they could not be hitting it with radiation or trying to find its molecular nature.

But in order to qualify as a material entity, more is required than that something be an object or a target for X-rays. It must evolve, develop, enter into reciprocal interactions with its surroundings.

(1) Weismann's theory postulated an immortal germplasm which could be reshuffled but neither created nor destroyed. The later mapping of the chromosome and the study of the recombination of different sections of the parental chromosomes in the offspring reinforced the idea that genetic differences among organisms can arise without altering the genetic material at all. And throughout the period of the debate, genetics did not consider the question of the origin or evolution of the gene. Therefore, Weismannian germplasm was, in its essence, anti-evolutionary. It allowed change, but only as a surface phenomenon, the reassortment of the unchanging entities. The Lysenkoist philosophers counterposed the Weismann – Morgan – Mendel school to Darwinism. And their more politically minded colleagues pointed out that scientific theories which deny the reality of change are generally associated with loyalty to the political *status quo*. Thus the metaphysical gene theory was also reactionary.

Mutations are of course changes in genes, but they are accidents or external and not part of the normal development of matter.

The rigidity of the gene concept was reinforced when the question of the origin of life was taken up seriously outside of communist circles, and was often reduced to the question of the origin of the gene.

(2) The relation between genotype and phenotype in genetics is a one-sided one in which genes determine phenotype but there is no reciprocal influence. Further, 'determine' is simply an evasion of what really happens in development. In the textbooks and in the practice of most geneticists, genetic determination carried with it an aura of fate.

The role of environment in the determination of phenotype was of course acknowledged, but in a subordinate way: 'The genes determine the potential, the environment its realization. The genotype is the size of the bucket, the phenotype is how much of it is filled.' Statistical techniques around the notion of heritability attempted to partition phenotype into hereditary and environmental components, but still as separable entities. Among Lysenko's adversaries, Schmalhausen[8] in the Soviet Union and Dobzhansky[9] in the United States were almost alone in emphasising a more sophisticated view of genotype – environment interaction in which the genotype is the norm of reaction to the environment. The subsequent development of the whole field of adaptive strategy is derived from their approach.

The one-way relation between gene and environment also emphasised the contradiction in genetics that all cells are supposed to have the same genes, yet produce different tissues.

(3) *Structure and process.* Western science as a whole is structuralist. That is, processes are seen as the epiphenomena of structures. Heredity implies an organ of heredity, memory implies an organ of memory or engram, language implies an archetypal capacity for language. In contrast, Lysenko's dialecticians emphasised process as prior to structure and saw structure as the transitory appearance of process. To them it was as absurd to look for the organ of heredity as for the organ of life. Heredity is a dynamic process in which various structures may be involved (Lysenko acknowledged the existence of chromosomes, assumed they had some function or other, but did not seem to consider it important to find out what that function was), and the model for it is metabolism, the exchange and transformation of substances between organism and environment.

(4) *The role of chance.* Ideas of chance play an important role in two aspects of genetics. First, the laws of Mendel and of Morgan are couched in terms of probability. Given the genotype of the parents, it is not possible to predict the genotype of an offspring exactly, but only to describe the distribution of genotypes in a hypothetical infinitely large *ensemble* of offspring. It is possible to *exclude* some genotypes, but in general there is no certainty about which of the possible genotypes an offspring will have. For characters of size, shape, behaviour, and so on, this uncertainty is further compounded by the variable relationship between genotype and phenotype. Second, mutation is said to be random, by which is meant that mutagenic agents like X-rays do not produce a single kind of mutational change in every treated individual, but rather a variety of possible mutations with different frequencies. The same uncertainty exists with respect to so-called 'spontaneous' mutations which appear unpredictably in individuals and are of many different types.

For Lysenkoists, these notions of chance seemed anti-materialist, for they appeared to postulate effects without causes. If there is really a material connection between a mutagenic agent and the mutation it causes, then, *in principle*, individual mutations must be predictable and the geneticists' claim of unpredictability is simply an expression of their ignorance. To propose that chance is an *ontological* property of events is anathema to Marxist philosophy.

The response of most geneticists, and certainly those of the 1930s, was that the unpredictability in genetic theory was *epistemological* only. That is, geneticists would agree that there was an unbroken causal chain between parent and offspring and between mutagen and mutation, but that the causal events were at a microscopic or molecular level not accessible in practice to observation, and not interesting to the geneticist anyway. It was their contention that, for all practical purposes, mutations and segregations are chance events. More recently, geneticists, invoking principles of quantum mechanics, make the stronger claim that the uncertainty of mutation is an ontological uncertainty as well, and here they come into direct conflict with the whole trend of Marxist philosophy. That issue, however, far transcends questions of genetics.

THE CONDITIONS CREATING LYSENKOISM

The books of Medvedev and Joravsky show clearly the way in which dogmatism, authoritarianism and abuse of state power can help to propagate and sustain an erroneous doctrine and even establish its primacy for a time. But a theory of 'bossism' is not sufficient to explain the rise of a movement with wide support nor to explain its form and context.

There were a number of streams that converged to give rise to and sustain the Lysenkoist movement. These were: (1) the material conditions of agricultural production in the Soviet Union; (2) the problems of agricultural experimentation under these material conditions; (3) the actual state of genetical theory and practice in the 1930s; (4) the ideological and social implications drawn from Mendelism, including the eugenics movement; (5) the response of the peasants to the collectivisation programme beginning in 1929; (6) the class origins of agronomists and academic scientists in the decades after the revolution and the strong cultural revolutionary movement toward popularisation of scientific understanding and activity; and (7) the growing xenophobia of the 1930s.

C

(1) *Conditions of Agriculture*

There can be no understanding of Lysenkoism that does not begin with
the hard facts of climate and soil in the Soviet Union. Since it is usual
both within and without the Soviet Union to compare Soviet and
American agricultural production, the same comparison of geography
is illuminating.

Nearly all of the Soviet Union lies above the latitude of St Paul,
Minnesota (40° N) so that its general temperature regime is more like
that of Western Canada than the United States. The growing season in
the most productive, *chernozem*, belt is short, and contrast between
summer and winter temperature is extreme, as compared with Western
Europe and the United States. Table 2.1 makes the point clearly, showing the dramatic increase in 'continentality' of the climate moving from
West to East in Europe and Asia along the 50° parallel of latitude. While

TABLE 2.1

	Number of frost free days	°C difference between warmest and coldest month
Utrecht	196	16·4
Berlin	193	19·3
Kiev	172	25·3
Kharkov	161	28·3
Saratov	151	30·6
Orenberg	147	37·4
Akmolinsk	129	37·3
Irkutsk	95	38·1
Pierre, South Dakota	161	32·6
Hutchinson, Kansas	182	27·8
Ames, Iowa	159	30·4

the Soviet Union has one-third more population than the United States,
acreage harvestable per year is the same, about 360 million acres. The
rich black *chernozem* soils of the Soviet Union, equivalent to the Great
Plains and prairies of the United States and Canada, are in a narrow
east-west belt from the Ukraine in the west passing just north of the
Black Sea to Akmolinsk in the east, running roughly along the 50th
parallel. South of this *chernozem* belt, rainfall is ten inches or less per

year and so is much too arid for normal agriculture. North of the *chernozem* belt, rainfall is sixteen to twenty-eight inches per year, quite adequate for agriculture, but the soil is poor, and the growing season short, and the winter frosts very severe so that neither winter nor spring wheat is favoured. The general problem in this region is to plant late enough to avoid killing frosts, yet early enough to get a full growing season. The *chernozem* belt itself, which is the chief agricultural region of the Soviet Union, lies in a band of marginal rainfall of ten to twenty inches per year with frequent droughts that result in catastrophic crop failures. In contrast, the United States' black soil belt runs north-south in the Great Plains, spans a broad range of temperature regimes, mostly milder than in the Soviet Union, and receives fifteen to twenty-five inches of rain per year, reaching thirty inches in the eastern-most sections. In addition, the United States has a large central section, just east of the plains, with thirty to forty inches of rain, three to ten feet soils, a long and mild growing season with summer nights not falling below $55°$ F, that is ideal for maize. This 'corn belt', which is the basis for meat production, is completely absent in the Soviet Union.

Lysenko's rejection of hybrid corn and his insistence on the use of locally adapted varieties is usually offered as a prime example of the counter-productive effects of his unscientific theories, while Khruschev is praised for his adoption of American hybrid corn breeding. Yet hybrid corn has not been a success in the Soviet Union precisely because of the lack of a 'corn belt'. In the United States, outside of the corn belt, in more marginal areas for maize, locally adapted varieties commonly out-perform hybrids.

The generally poor conditions for food crops is matched in other cases. Cotton, which is chiefly produced in the moist warm regions of the south-east of the United States by dry farming, must be irrigated, at considerable expense, in the Soviet Union, since, there, warm temperatures are accompanied by semi-aridity.

The most striking example of the deleterious effect of environment on a staple crop is for sugar-beet, the standard sugar source in Europe. In Germany and France, with high summer rainfall, yields in the mid-1930s were about thirteen tons per acre, of which 34 per cent was sugar content. In the Soviet Union, with dry hot summers, yields were four tons per acre with a sugar content of only 27 per cent.

Another problem for Soviet agriculture is that much of the arable land cannot be cropped annually, nor can it be planted with high yielding varieties that remove moisture and nutrients from the soil at a high rate. For example, 45 million acres in Kazakhstan can be cropped only every second or third year. Soviet agriculture must then be more extensive and less intensive than American both in space and time, although both are non-intensive in comparison with most European practice, but for different reasons. Soviet agriculture is extensive because of the generally severe conditions of climate and soil, while the Americans have sufficient favourable climate and land to make intensive agriculture unnecessary and unprofitable.

Thus, the figures for important food crops shown in Table 2.2, taken from the 1930-5 data, shows the intensive agriculture of Western Europe in sharp contrast to the American and Soviet practice. The yields for the Soviet Union are over-estimates by perhaps as much as 20 per cent since they were normally estimated in the field rather than actually measured after harvest.

TABLE 2.2

Yields in bushels per acre 1930-5[10]

Crop	Germany	France	United States	Soviet Union
Wheat	29·7	23·0	13·5	10·8
Rye	27·4	18·3	10·7	13·5
Barley	35·9	26·6	20·1	16·0
Corn	–	–	22·1	16·3
Potatoes	226	164	108	120

In general, Soviet agriculture is carried out in conditions that are not only marginal *on the average*, but of much greater *temporal uncertainty and variation*. Periodic catastrophes from drought or severe winter frosts are a regular feature. Two successive years of drought in 1920 and 1921, coming hard on the heels of the Civil War, caused a catastrophic famine in which more than a million perished. Again, in 1924 there was a very severe year which reduced grain supplies by 20 per cent. This variability and unreliability of temperature and rainfall and the imminent possibility of agricultural catastrophe must be regarded as the leading element driving Soviet farm policy. It is no accident that the first wholesale trials of vernalisation[11] were carried out after the two severe winters of 1927-8 and 1928-9 in which 32 million acres of winter wheat were lost in the extraordinary cold.

(2) Problems of Experiment and Evaluation

The normal American method of variety testing is to plant a number of varieties in several years at several locations and choose those varieties with the highest average yield over locations and years, with some attention paid to variation between years and locations when average yields are very close. The underlying model is of normal variation around a mean, the coefficient of variation being fairly low so that any sequence of a few years averaged over a few localities will not deviate greatly from any other sequence. This is, in fact, the model that underlies all normal statistical analysis of experimental science. Events are assumed to be regular and drawn from a 'homogeneous' distribution. But the experience of Soviet agriculture has been quite different. There is generally a sequence of 'normal' years punctuated at uncertain intervals by one or more severe crashes. While years and locations could be averaged, the value of such averages as predictors is poor, because the coefficients of variation are so high. An analogy from ordinary experimental science makes the distinction clearer. When a new experimental technique is worked out, there will be a period during which the experimenter has such poor conditions that there will be some replications of the experiment that are clearly deviant and not regarded as part of the normal experimental variation. Not until the experimenter has the system under sufficient control to avoid these deviant cases will data begin to be accumulated to test some hypothesis. The decision that the system has passed from the initial uncontrolled stage of heterogeneous results to the stage of controlled variation is made impressionistically and represents a change in the underlying model of the universe with which he is dealing. In the first stage, averages over all experiments are not appropriate, and, forced to characterise the results, the experimenters would perform some culling, averaging only the 'normal' replications which represent the 'potential' of the experiment.

This is precisely the procedure followed by the Lysenkoists, and by Soviet agriculture authorities even before the Lysenkoist movement, when they reported yields per acre. Obviously, such a culling procedure can be and has been used for self-serving purposes since there is no objective way to decide which cases are 'deviant' and which are 'normal'. That this 'pathological' model played into the hands of unscrupulous manipulation, or was unconsciously used by wishful thinkers, cannot be doubted. But the conventional statistician's scornful demand that *all* the data be averaged in an 'objective' way will not serve either. The immense variation in results make the averages meaningless as predictors.

Lysenkoist recommendations had such wide appeal precisely because they were intended to cope with extreme environments. Vernalisation, for example, was designed to avoid winter killing of wheat, while 'sowing in the mud'[12] was designed to give plants a very early start against the summer droughts. It is revealing that the report on vernalisation of a 1931 drought conference carried a 'warning against drawing hasty negative conclusions from possible individual failures', because 'particular failures are possible, indeed unavoidable . . . as in every experimental search for new pathways.[13] Apparently it was the hope of the conferees that these 'experimental pathways' would soon come under sufficient control to avoid the 'particular failures'.

Normal procedures of variety testing and normal statistical evaluation in which equal weight was given to all observations could not have been applied successfully in the conditions of Soviet agriculture of the 1930s because the level of agricultural technology and husbandry was insufficient to buffer against the extremes of climatic variation. It is not certain that even today conventional plant breeding and evaluation techniques could be successful. What is required is some objective method of dealing with the uncertainties. Perhaps the concepts of *maximum* and *maximax* solutions to the 'game against a capricious nature' could be used, although the irony would be great since games theory is a unique development of bourgeois economics.

(3)   *The State of Genetical Theory*

The Lamarckian theory — that characteristics acquired by the organism as a response to the environment during its lifetime may be transmitted to its offspring — had never really been refuted so much as it had been abandoned with the development of modern genetics. The textbook refutation of Lamarck was the work of Weismann.[14] In his classical experiment with mice, the removal of the tail over successive generations failed to produce mice with shorter tails. However, this was in fact irrelevant to the Lamarckian hypothesis, which never claimed that mutilations are heritable. Rather, the claim was that active adaptive responses are transmitted to the offspring; and in support of this there was an impressive body of experimental data.

Among the classical Lamarckian experiments were those of Guyer, who found eye defects in the offspring of rabbits injected with corneal antibodies; the work of Jollos on the transmission of heat resistance and other traits induced by heat treatment in *Drosophila*, Cunningham's arguments on the evolution of the beehive; and MacDougall's behavioural experiments. In plants, Lucien Daniel studied graft hybridisation and Lesage adapted cress to particular conditions and claimed the transmission of the adaptation over six or more generations. Bolley, working with flax in North Dakota, claimed to induce disease resistance which is transmitted. About 1939, Eyster described experiments in which corn was grown in different parts of the United States. The kernels showed different colour patterns, and 'under California conditions more of the color changes extended into the germplasm and this became genetic'. As late as 1944, Reynolds published in the *Proceedings of the Royal Society* a paper entitled 'On the Inheritance of Food Effects in a Flour Beetle, *Tribolium Destructor*' in which the feeding of thyroid extract had a greater effect in the next generation than on the animals fed the thyroid.[15]

Weismann's argument was not based merely on his negative experimental results. Prior to the rediscovery of Mendel's laws in 1900 he had already formulated the distinction between germplasm (or hereditary material) and somatoplasm (the rest of the body) and argued the impossibility of the inheritance of required characters on the basis of the anatomical separation of the two early in development. Reviewing the embryological argument in 1948, Berrill and Liu concluded:

> There is little doubt that he [Weismann] read into his observations ideas that were in a sense already in the air . . . . But it is primarily on the basis of strict recapitulation that Weismann propounded the migration of the primordial germ cells, to which he so stubbornly adhered that he seemed to have defended it to the extent of disregarding the truth. His interpretation of the germ cell origin of *Coryne* serves to illustrate how far imagination can be pushed to suit a preconceived idea . . . . The weight of authority, however, of the Weismann – Nussbaum combination convinced many later workers of the existence of facts they could not observe.[16]

A special form of the inheritance of acquired characters is graft hybridisation, in which grafted plants acquire and supposedly transmit some of the characteristics of their graft partner. This phenomenon likewise has a long pre-Lysenko tradition. In the *Cyclopedia of American Horticulture* (New York: Gordon Press, 1974) Liberty Hyde Bailey discusses the uses of graftage in plant propagation, and adds:

> There are certain cases, however, in which the scion seems to partake of the nature of the stock; and others in which the stock partakes of

the nature of the scion. There are recorded instances of a distinct change in the flavor of fruit when the scion is put upon stock which bears fruit of a very different character. The researches of Daniel (1898) show that the stock may have a specific influence on the scion, and that the resulting [changes] may be hereditary in the seedlings.

Thus, when Lysenko and his followers began to put forward claims of directing hereditary change in the 1930s, Lamarckism was not a dead relic dredged up from the past. It had already been rejected almost universally among geneticists, but was still very much alive in palaeontology and horticulture, and had an extensive literature of experimental results which had never been adequately refuted.

Geneticists were largely unaware of, or indifferent to, the Lamarckian tradition. They regarded it as a carry over of pre-scientific folk science. And in so far as they confronted Lamarckism at all, it was to reject it out of hand because the organisms used were not well characterised, the characteristics supposedly modified were not clear-cut phenotypes like the fruit fly mutants they favoured and the research reports were especially deficient in statistical sophistication. They assumed that Lamarckian results could be explained by hidden selection processes. And, in any case, the impressive successes of Mendelian genetics and the chromosome theory made it simply unnecessary to consider vague allusions to physiological interactions in explanation of dubious claims by not quite respectable authors. (The academic community is as quick as any small town to declare someone a crackpot and not quite believable. The disabilities attached to such a judgement may be anything from smirks to difficulties getting published, even greater difficulties getting read, to unemployment. This is made easier if the person in question lacks formal academic credentials, such as the early twentieth century plant breeders, Burbank or Michurin in the United States and Soviet Union respectively but also applies to wayward colleagues. Thus a whole scientific community may be personally aware and yet intellectually unaware of dissident currents.)

Meanwhile, genetics itself was changing. New phenomena were appearing which were difficult to assimilate. There were the 'dauer-modifications,' changes induced in lower organisms which were transmitted in diminishing degree over as many as twenty generations. New kinds of material and extra-chromosomal inheritance were being described. Hereditary particles outside the nucleus ('plasmagenes') were postulated, and hints as to the special role of the nucleic acids in heredity were appearing. The Lysenkoists watched this literature very closely. They saw in it signs of a general crisis in genetics in which *ad hoc* hypotheses and ignored data presaged the final fall of the gene theory.

The contrasting reactions of geneticists and Lysenkoists to the Griffith experiments[17] show how two opposing paradigms can respond so differently to the same experience and each emerge reinforced. The experiment was as follows: a number of different strains of the pneumococcus bacteria exist which can be distinguished by their virulence or non-virulence and by whether the outer capsule is present or absent. Griffith found that live pneumococcus of one variety acquired some of the characteristics of dead bacteria of another strain injected into the same host animal.

From the point of view of genetics, this was an important step in the identification of the genetic material as nucleic acid. From the Lysenkoist point of view, the heredity of one strain of bacteria was transformed by exposure to a specific environment, namely killed bacteria of the other strain. It is therefore by definition the inheritance of an acquired character and the experiment was widely quoted. The important point is that they were formally correct, and that for them this formal precision completely obscured the scientific significance of the experiments. This same approach characterised their treatment of the other anomalies of genetics and cytology. Mendelian genetics asserts that the nucleus controls heredity. But the so-called 'plasmagenes' refuted this. Chromosomes are supposed to be linear arrays of genes, but the best microphotographs of chromosomes showed a distinctly non-linear structure with thousands of loops coming off the chromosomes in a so-called 'lamp-brush' structure.

All of the scientific possibilities opened up by newly discovered phenomena were obscured by a legalistic 'is this or is it not the inheritance of acquired characters?' 'Does this or does it not show extra-nuclear inheritance?' 'Is genetic change directable or not?'

(4) *The Ideological and Social Implications of Genetics*

It is essential to distinguish between what we might call the 'minimal theoretical structure' of a science, which is dependent upon unspoken ideological assumptions, and a kind of ideological superstructure that is built upon but is not logically entailed by the minimal structure. For Mendelian genetics, the minimal structure includes the laws of Mendel and the Weismannian principle that the material substance whose behaviour is formally described by the Mendelian laws cannot be altered in a directed and adapted way by information from the environment, but that the phenotype of an organism is the outcome of the biosynthetic activities of genes taking place in a particular sequence of external and internal conditions. The ideological superstructure that has been laid on this theory by various geneticists includes notions of the 'limits' set to the phenotype by the genetic 'potential', the notion that what is inherited is somehow fixed and unchangeable, that organisms are 'determined' by their genes. By acting as if this ideological superstructure were in fact, the substance of genetics, geneticists invite a misplaced quarrel with the minimal structure itself. In 1931, Zavadovsky[18] foresaw the inevitable attack on Mendelian genetics that was being invited by the biological and genetic determinism and the pernicious eugenic elitism being read into their science by geneticists. He warned against the extreme environmentalist counter-reaction that would attempt to destroy all of genetics in order to assert the plasticity and perfectability of human society. He was the first, as far as we know, to point out that Lamarckism was anti-progressive since it would imply that centuries of degradation and brutalisation of workers and peasants had made them genetically inferior as well.

Among the issues of superstructure were several prominent ones. In the mid-1920s, most Soviet and Western geneticists propagated an elitist and racist eugenic ideology,* discussing for instance the breeding of superior types from the ranks of the intelligentsia as well as from those members of the lower classes who had been in the vanguard of the revolution.[19] Eugenicists also claimed that the genetically 'best' elements in the population were being outbred by the 'worst' and that this trend might grow worse with population control. This kind of naive genetic determinism of human behaviour naturally invited ideological attack.

The treatment of the gene merely as a cipher, a book-keeping device, uncoupled genetics from physiology. Thus Bateson explained the Mendelian view to the New York Horticultural Congress in 1902 roughly as follows: 'the organism is a collection of traits. We can pull out yellowness and plug in greenness, pull out tallness and plug in dwarfness.' This uncoupling, so attractive to geneticists and to Anglo – American analytical reductionism, was offensive to Lysenko's group who saw heredity as a special (but not too special) case of physiology.

---

*Rose and Hanmer suggest that the proposals were also sexist. See 'Women's Liberation, Reproduction and the Technological Fix' in the first volume, *The Political Economy of Science*.

Mendelian genetics, which made the possibilities of artificial selection depend on the fortunate occurrence of useful genes, a small minority of the mutants, imposed limits to the progress of plant breeding which were socially unacceptable given the needs of Soviet agriculture. On the other hand, a model in which the creation of hereditary variation proceeds apace with its selection promised indefinite progress once physiological knowledge was sufficiently sophisticated.

The traits used by Mendelian genetics to develop and argue its theory are clear-cut mutants of *Drosophila* (fruit fly) and a few other organisms. These mutants are a special kind of variant. They are usually non-viable in nature. They were chosen for their unvarying expression so that they could be followed easily while the complicated processes of variable expression so common to adaptive, quantitative and agronomically important traits were ignored. Finally, many of the mutants and chromosomal abnormalities were artificially induced by radiation at dosages so far beyond those that occur in nature as to make it appear that Mendelian genetics dealt with a special class of laboratory phenomena but could not, in principle, deal with the problems such as adapting fruit trees to the far north.

(5) *The Reaction of the Peasants to Collectivisation*

Unlike the Chinese revolution, with its strong political base among the peasants, the Bolshevik revolution could not count on a political and revolutionary peasantry, although 80 per cent of the population was rural. Thus, while Chinese agriculture is rapidly passing from co-operative to collective chiefly by persuasion and local voluntarism, the Russian peasantry, steeped in a petty bourgeois notion of eventual individual land ownership and encouraged in that concept by the market economy of the New Economic Policy, was totally unprepared for the collectivisation that was required by a rational socialist economy. For the Russian and Ukrainian peasants, collectivisation meant appropriation of land and agricultural products by the urban population. It was all one to the peasants whether the product of their labour was taken by a landlord or by a revolutionary government. After all, it was not *their* revolution.

The pressing demand to feed the urban working population resulted in a pace of collectivisation far in excess of what could be supported by the political state of the countryside. So, when in 1928 the wholesale collectivisation of agriculture began without the long and difficult task of revolutionising the peasants having been accomplished, it was met by forceful resistance and sabotage. Agricultural production was wrecked by ploughing under crops, refusal to sow and harvest, the wholesale slaughter of livestock and attacks on agricultural officials. This force was met with greater and more terrible state force, which eventually won the day for collectivisation but at a great cost in lives, material wealth and political development. Crop yields in 1929 – 30 were 15 – 20 per cent below the pre-collectivisation figures and much more below the optimistic projections of the first five-year plan.[20] Hostile writers like Joravsky and Jasny lay the blame for these losses on the programme of collectivisation, rather than on the peasants' use of force and sabotage to protect their private property. This point of view blinds these authors to the reality of the 'wrecking' and 'sabotage' (which they always put in inverted commas) that characterised Soviet agriculture at the end of the 1920s and in the 1930s. It is entirely reasonable that charges of 'wrecking' levelled by Lysenkoists against their opponents, as an explanation of the failure of proposed methods, should be believed by agricultural officials. An atmosphere of hostility and distrust, grounded in bitter experience, permeated the relations between the state agricultural organs and the mass of farmers. We come here to another aspect of the normal – abnormal model of production discussed in relation to climate. The very real sabotage of agricultural production led to suspicion that instances of failure of Michurinist methods, which, after all, could show striking successes in *some* years and *some* localities, must be the result of abnormal conditions created by the wilful resistance of saboteurs among farmers and agricultural scientists.

(6) *The Class Origins of Scientists and Agronomists*

The suspicion of the more academic 'pure' scientists, including most geneticists, arose in part from their actual histories. Most of the senior scientists of 1930 had been members of the intellectual middle classes of pre-revolutionary Russia. Many had favoured the February revolution but had strongly opposed the Bolsheviks. Men like Vavilov, who was enthusiastic about the socialist revolution from its early days and who displayed a great enthusiasm for the possibilities of science and agriculture in the new society, were the exception. Nevertheless, most agricultural specialists and scientists were kept on in responsible positions because the state seemed to have no choice. Not only in science, but in all branches of technology and management, unsympathetic managers and technicians had to be employed in socialist enterprises if there was not to be a complete breakdown. Soviet authorities were conscious of the difficulties of such a procedure and the position of such pre-revolutionary holdovers was problematical.[20]

In contrast, Lysenko represented the Russian equivalent of the 'horse-back plant breeder', coming from peasant origins and receiving the bulk of his technical training after the revolution. Over and over again the polemic of Lysenkoist and anti-Lysenkoist contrasts are the 'priests' of 'aristocratic and lily-fingered' science with the 'muzhik's son' who is 'illiterate' and 'ungrammatical'. This contest between the effete middle-class intellectuals, and the close-to-the-soil practical agronomists was subtly extended to include a conflict between theory and practice, a vulgarisation of Marxism. In every aspect the conflict in agriculture was a revolutionary conflict, posing the detached, élite, theoretical, pure scientific, educated values of the old middle classes against the engaged, enthusiastic, practical, applied, self-taught values of the new holders of power. That is why Lysenkoism was an attempt at a cultural revolution and not simply an 'affair'.

One of the elements of the cultural revolution was the terror. Joravsky, after a thorough analysis, concludes that: 'Anyway one searches it, the public record simply will not support the common belief that the apparatus of terror consciously and consistently worked with the Lysenkoites to promote their cause.' He points out that the general class divisions between geneticists and Lysenkoists would, in any event, result in more geneticists suffering under a revolutionary terror. While that is undoubtedly true, it must also be the case that the existence of a revolutionary terror, the preponderance of Lysenkoists among state officials, and occasional veiled suggestions by Lysenkoists that they did have access to the organs of terror, would be quite sufficient to inhibit the overt activities of geneticists. Speculation on the way the revolutionary terror might have operated if there had been no historical and class divisions between Lysenkoists and geneticists really misses the point that the struggle *was* in large part a class conflict.

A dispute among plant breeders and geneticists does not invariably become a national *cause célèbre*. However, under Soviet conditions of the 1930s, it quickly became a public issue. One of the early achievements of the Soviet regime was mass publishing. Long before the paperbacks became a lucrative business in the United States, the Soviet Union was publishing world classics, scientific works, poetry and political tracts in cheap editions of tens or hundreds of thousands. The ubiquity of bookstores is a striking feature of socialist cities the world over.

Within this general literacy, science played a special role. There was a widespread consciousness of the relative backwardness of the country and the urgency of rapid technological advance based on the development of science. The development of the Academies of Science of the non-Russian Republics was considered a major step in liquidating the cultural vestiges of Czarist colonialism in central Asia and the Caucasus. This interest in science merged with the older traditional socialist belief that through a scientific understanding of the world, it can be changed for the better — a belief which made at least evolution and cosmology a part of the general liberal education of socialist workers, and before that led Engels into essays in mathematics, tidal friction, human evolution and cosmology.

The Soviet cultural interest in science was especially excited by the broadest, large-scale theories. Vernadsky's concept of the biosphere, Sukachev's biocoenosis (which attempted to treat whole systems, such as forests), Vasili Williams' soil science, which treated the soil as a living system in co-evolution with its vegetation and agricultural practice, Oparin's opening up of the origin of life, Pavlov's exploration of the organisation of behaviour, were both intellectually exciting and aesthetically appealing.

The general alertness to, and interest in, science was heightened by the special practical concern with agriculture and the Soviet food supply.

Here Lysenko had a decided advantage. He was on the offensive, promising advances where geneticists advocated caution. He mobilised large numbers of farmer-innovators whose exploits in plots on collective farms were publicised along with the Stakhanovite innovators in industry. The excitement of bold, sweeping theories, popular inventiveness, the rejection of academic-elitist stodginess in the face of novelty, defiance of the received wisdom, all created an exuberant cockiness described some years earlier by Stalin in his pamphlet *Dizzy with Success*. It was the exultation in the achievements of the early years of the revolution that led to a sense of omnipotence, of daring to do the impossible, of intolerance toward doubters which Stalin was able to perceive, describe, and denounce, but not quite resist.

(7) *Xenophobia*

The Lysenkoist distrust of established academic authority included both Soviet and foreign geneticists, and was originally part of the iconoclastic exuberance and anti-elitism shared with other sections of Soviet society. But as political and philosophic issues became more prominent in the debate, foreign science was increasingly seen as hostile, as part of the capitalist encirclement. On the naive assumption of a simple one-to-one relation between someone's views in genetics and his or her general political outlook, the anti-Soviet or racist attitudes of foreign geneticists were used to discredit the science. Then sympathy with their scientific views was increasingly assumed to imply political sympathy as well, and close intellectual ties of Soviet with foreign scientists was taken to justify suspicion of disloyalty.

Thus, within a short time, the healthy demand for Soviet intellectual independence was converted into a grotesque xenophobia. It was through this route that Lysenko's opponents were subject to political suppression. The most notorious episode was the arrest of Nikolai I. Vavilov in August 1940. Vavilov, a pioneer in plant genetics and the evolution of cultivated plants, was seized while on a field trip in the western Ukraine and charged with wrecking activities. The particulars included belonging to a rightist conspiracy, spying for England, leadership in the Labour Peasant Party, sabotage in agriculture, and links with anti-Soviet *émigrés*. He was sentenced to death by a military court, and although this was later commuted to ten years' imprisonment, Vavilov died in prison in 1943.

While from the point of view of the Lysenkoists the charges of disloyalty removed leading opponents and silenced other critics, from the viewpoint of the police apparatus their victims' scientific views and international contacts were merely evidence of anti-Soviet activities. Intellectual wrecking — the deliberate making of wrong decisions for the purpose of sabotage — was a respectable accusation in the Soviet Union. In the early 1930s, several British engineers had been convicted, apparently justly, of sabotaging some of the projects of the first five-year plan. Later, in the major purge trials, physicians were accused falsely of murdering the writer Maxim Gorky by deliberate prescribing treatments that would weaken his lungs, already the weak point in his body. This tradition was continued into the post-war period in the infamous doctors' case.

It would not be correct to interpret the anti-foreign hysteria of the late pre-war and early post-war periods as a simple revival of Russian nationalism. Rather, it represented a new, typically socialist form of xenophobia derived from a distorted appreciation of real problems. Scientists in the newly post-colonial countries are very aware of the need for intellectual independence. They recognise that the Western hegemony of science is an instrument of domination. They are aware of the dangers of an excessive regard for established centres of science which leads to the illegitimate transfer of techniques, reinforces the hierarchical, elitist social structure of science, and fosters the ideology of neutral technocracy. In this context, the lesson of socialist xeno-phobia is not that socialist scientists should return to the fold of the international (largely bourgeois) community of science as the only alternative to a Lysenkoist rampage. Rather, it leads to the demand for a programme of active evaluation and selection of those aspects of foreign science which can be incorporated into the construction of socialist science and a militant resistance to scientific colonialism. This requires a total rejection of the simplistic bureaucratic dogmatic Marxism which sees only the unity of phenomena and therefore equates the philosophy, scientific content, social context and political ideology in foreign science without seeing the heterogeneity and contradictions in it. Ideologically, it means a reaffirmation of dialectical analysis, and this in turn depends on free discussion without administrative fiat.

THE APOGEE AND DECLINE OF LYSENKOISM

While in 1940 there was still lively debate on the genetics question in the Soviet Union, by 1948 Lysenko had won the official backing of the party and ministries. Some of his opponents lost their positions. Others pretended to go along with him and continued at their institutes. Some transferred to the biophysics programmes under the protection of the Institute of Physics. A few, like Schmalhausen, conducted a spirited rearguard defence of genetics.

What happened in the interim was of course the war, reconstruction and the Cold War. In 1946, Churchill announced the Cold War in his Fulton, Missouri speech. In 1947 the Cominform (Communist Informa-tion Bureau) was organised to replace the defunct Comintern, and Andrei Zhdanov put forth his thesis of the world divided into two camps. The communists were driven out of the post-war coalition governments in France and Italy. By 1949, the North Atlantic Treaty had been signed, the first of the network of US-dominated alliances encircling the socialist world.

Effective intellectual contact between the Lysenkoists and genetics all but ceased. A few of Lysenko's supporters attended international genetics congresses, but Soviet anti-Lysenkoists did not appear even when they were on the programme. The genetics congresses deplored their absence, while plans had been made to urge their defection and offer them jobs in the West if they had come.

Meanwhile, most of the Lysenkoist work was ignored in the capitalist countries, where, aside from occasionally quoting absurd claims for ridicule, interest centred on the administrative abuses of an aggressive Lysenkoism backed by the Soviet Communist Party. The disinterest in the scientific side of the dispute was such that when, in 1948, an advertisement in *Science* offered for sale translations of several of the best Lysenkoist research papers, there were only eight responses. In the context of the Cold War, even the suggestion that Lysenko's work ought to be examined cost Ralph Spitzer his position as a professor of chemistry at Oregon State University.

The very limited contact between Lysenkoism and genetics was through anti-Lysenko Soviet and East European geneticists and Western scientists who were either themselves pro-communist or at least not blinded by the hysterical anti-communism of the times to the extent of refusing to examine the claims.

Schmalhausen in the Soviet Union and Waddington in Britain were finally able to show the basis of the apparent inheritance of acquired characters through the discovery of genetic assimilation, the process whereby latent genetic differences within populations are revealed but not created by environmental treatment, and therefore become available for selection. Scattered researchers in Japan, France, Switzerland, Britain and the United States repeated some of the experiments of Lysenko's group. But these were the exceptions.

In several Western countries, leading biologists were effectively driven out of the communist parties because of their opposition either to Lysenko or their party's endorsement of Lysenko. Thus another possible channel of communication was cut off. In this context of the Cold War and the 'two camps' doctrine, Lysenkoism became more strident, politically opportunist, more reckless in its claims. Whereas earlier Lysenkoism emphasis that it is not at all easy to modify the the heredity of organisms, and that responses to the environment are often barely perceptible, later Lysenkoists claimed to transform wheat into rye in a single step. Lysenkoists were never as ignorant of Western genetics as their counterparts were of Lysenkoist work. However, their use of this literature was mostly to search for 'admissions' – admissions of the incompleteness of genetic theory, of our understanding of chromosome behaviour, of possible cases of extra-nuclear inheritance, and so on. For example, Prezent quoted Franz Schrader, the American cytologist, as admitting that 'in the cytology of *Drosophila* itself there is much that does not conform to what we have set up as the standard course of events'. This search for gaps, admissions, ambiguities, symptoms of a crisis in genetics, had something of the spirit of a Jehovah's Witnesses' tract on evolution in which palæontologists' comments on gaps in the fossil record were taken as evidence that the whole theory was false, and that at least its more perceptive practitioners recognised this.

This approach, which we interpret as a crude, simplistic interpretation of the 'two camps' doctrine, according to which socialist science had to reject and overthrow bourgeois science lock, stock and barrel, made it extremely difficult for Soviet biologists to respond to new phenomena in genetics. All results were read as either still holding to the Morgan – Mendel doctrine or as a tentative departure from it.

The experimental refutation or reinterpretation of Lysenkoist results probably had very little to do with the decline of Lysenkoism. As long as it maintained its institutional, administrative and ideological coherence, it could filter out disturbing arguments or evidence, assimilate the results of genetics into its own structure, and remain intact. A paradigm has a semi-permeable boundary.

The decline of Lysenkoism was accelerated by the development of modern genetics only after it lost its protective boundary.

(1) It did not fulfil its promises to Soviet agriculture. Agriculture remained the critical issue in the economy through the Khruschev administration and beyond. But the same cause that contributed to the rise of Lysenko in the 1930s had an opposite effect now.

(2) Economic planning and administration had meanwhile become increasingly depoliticised, the domain of experts and technicians. The slogans now were not so much revolutionary innovation but 'business-like' efficiency, cost accounting, balance sheets; the goal was not to develop an alternative, socialist technology but to adopt the most advanced American methods. This change was symbolised by Khruschev's visit to the Garst farm in the US corn belt.

(3) The incipient cultural revolution of the anti-elitist, populist element of the era of Stakhanovites and peasant innovators was aborted and the prestige of academic authority was reconsolidated. Perhaps for this reason, Lysenkoism has retained an attractiveness for those countries which are actively fighting the battle against the elite academy. Lysenko's administrative repressiveness has been rejected, but courses

in 'Darwinism –Michurinism' are still taught in some of the agricultural colleges of developing socialist countries and visiting lecturers are sometimes queried about Michurin's teachings. In some capitalist countries, certain Maoist sects are pro-Lysenko, some only vaguely, others with great firmness and conviction. So, for example, the Progressive Natural Sciences Study Group of Dublin reissued in 1970 a collection of articles by Lysenko illustrated by photographs of agriculture in China with captions such as 'In Socialist China the theories of T. D. Lysenko are constantly being put into practice', while a pamphlet of the Sussex Student Movement in 1971 described Lysenko as a

> Great upholder of materialist method of investigation and study in natural science, who firmly opposed all the unscientific methods of 'seeking' facts to prove preconceived notions in Biology that [are] still being promoted today. Because Lysenko upheld the scientific method of seeking truth from facts, he is now called by the scientific 'experts' a crank.*

(4) With the ebbing of the more raucous Cold War rhetoric and the development of an active coexistence approach, the 'two camps' model of science lost its appeal. Emphasis shifted to underlining the common ground and similarity of Soviet and American science. The sporadic warnings that coexistence in international politics did not imply coexistence in ideology were, at least in science, a futile rearguard action. The opposition of Lysenkoist and traditional genetics, previously a matter of pride, now became an embarrassment.

(5) The weakening of the political police power, the return of exiled geneticists, the urgency to settle accounts with the repressive aspects of previous administrations, coincided with the ideology of the specialists: the demand for freedom of scientific research not only from the imposition of ideological and political demands, but also from their influence.

DID LYSENKOISM AFFECT SOVIET AGRICULTURE?

It is commonly assumed that Lysenkoist techniques of agriculture and doctrines of biology had a serious effect on agricultural production. After all, if genetics is important for plant breeding, and plant breeding

---

*However, for the Chinese view on Lysenko, see the Introduction, Chapter 1 of *The Political Economy of Science* and also Science for the People, *China: Science Walks on Two Legs* (New York: Avon, 1974) pp. 125 – 6.

important for agricultural production, then such serious errors as were
propagated by the Lysenkoists must have been responsible for a disruption of progress in agricultural production. Yet what is the evidence for
such a disruption? Whatever the figures for agricultural output, it could
always be stated that they would have been higher if not for Lysenkoism. But the logic of such counter-factuals is not compelling and we
could as easily postulate that they might have been *lower* except for
Lysenkoism. What we can do, however, is to compare the history of
Soviet agricultural production before, during and after the predominance
of the Lysenkoists with the history of, say, American agriculture of the
same period. We then have both an internal comparison through time
and a cross comparison. Can we see in such comparison the postulated
negative effect of Lysenkoism? We have chosen to look at wheat yields
for this comparison since vernalisation of winter wheat was the first
Lysenkoist recommendation and one with which the movement came
to be identified. Indices of total agricultural production show the same
picture. Table 2.3 shows indices of wheat yields from 1926 to 1970 in
the Soviet Union and the United States.

The yields in the Soviet Union are over-estimated during the 1930s
by as much as 20 per cent, but the figures after the war do not suffer
from this problem, nor are the base years affected. The comparisons are

TABLE 2.3
Yields of wheat relative to base years 1926 – 8

	United States	Soviet Union
1926 – 8	100 (14·83 bu/acre)	100 (6·69 bu/acre)
1929 – 31	98	104
1932 – 4	82	93
1935 – 7	87	97
1938 – 40	96	113
1941 – 4	118	–
1945 – 7	118	72
1948 – 50	116	106
1951 – 3	116	135
1954 – 6	128	130
1957 – 9	159	172
1960 – 2	169	184
1963 – 5	175	162
1966 – 8	181	213
1969 – 70	207	236

remarkable. Both the US and Soviet productivity decreased during the 1930s, certainly for different reasons: in the United States because the depression caused a reduction in capital investment in agriculture, in the Soviet Union because of political problems associated with collectivisation, as well as problems of capital investment. During the war years, the Soviet Union suffered a catastrophic loss of productivity while it was recovering in the United States. Then, beginning in 1950, both countries began a period of rapidly increasing yields which kept pace with each other, the Soviet increases being somewhat higher. We should note that 1948–62, the period of Lysenkoist hegemony in Soviet agrobiology, actually corresponds to the period of most rapid growth in yields per acre! Moreover, even a time-delay hypothesis, supposing that the effects of Lysenkoism on genetical research are felt only later, is at variance with the observed continued growth in yields per acre. The data in the table are even more remarkable if it is noted that, during this period, the total acreage occupied by wheat increased in the Soviet Union from 30 million to nearly 70 million hectacres, while US acreage shrank from 60 to 45 million acres. Thus increased Soviet yields have been in spite of bringing large amounts of new and marginal land into cultivation, while the opposite process was going on in the United States.

While there may be particular crops and situations where Lysenkoist doctrines prevented the solution of some specific problems (breeding for disease resistance, perhaps) there is no evidence that Soviet agriculture was, in fact, damaged and Soviet yields have followed the same upward trend as yields in other advanced technologies, chiefly as the result of massive capitalisation of agriculture, including pesticides, fertilisers and farm machinery.

CAN THERE BE A MARXIST SCIENCE?

Lysenkoism is held up by bourgeois commentators as the supreme demonstration that conscious ideology cannot inform scientific practice and that 'ideology has no place in science'. On the other hand, some writers are even now maintaining a Lysenkoist position because they believe that the principles of dialectical materialism contradict the claims of genetics. Both of these claims stem from a vulgarisation of Marxist philosophy through delibrate hostility in the one case or ignorance in the other. There is nothing in Marx, Lenin or Mao that is or that can be in contradication with the particular physical facts and processes of a particular set of phenomena in the objective world.

The error of the Lysenkoist claim arises from attempting to apply a dialectical analysis of physical problems from the wrong end. Dialectical materialism is not, and has never been, a programmatic method for the solution of particular physical problems. Rather, dialectical analysis provides us with an overview and a set of warning signs against particular forms of dogmatism and narrowness of thought. It tells us: 'remember, that history may leave an important trace'; 'Remember that being and becoming are dual aspects of nature'; 'Remember that conditions change and that the conditions necessary to the initiation of some process may be destroyed by the process itself'; 'Remember to pay attention to real objects in space and time and not lose them utterly in idealised abstractions'; 'Remember that qualitative effects of context and interaction may be lost when phenomena are isolated'; and above all else, 'Remember that all the other *caveats* are only reminders and warning signs whose application to different circumstances of the real world is contingent.'

To attempt to do more and to try to distinguish competing theories of physical events, or to discredit a physical theory by contradiction, is a hopeless task. For every point of genetics that is supposedly contradicted by dialectical materialism, we can show that, in fact, there is complete support. To the Lysenkoist claim that Mendelism is idealist and formal, we respond that, on the contrary, Mendel solved the problem of heredity precisely by concentrating on the actual pattern of variation among the offspring of a cross, rather than trying to sum up the results in a single idealised description, as others did. Mendel's revolutionary insight was that variation was the thing-in-itself, and that by a study of the pattern of variation he could bring together the two apparently contradictory aspects of heredity and variation under a single explanatory mechanism. The synthesis of the two 'contradictory' elements, heredity and variation, by seeing them as dual aspects of the same phenomenon, was a triumph of dialectical thought. Of course there is a level of abstraction even in Mendel, and he took care to remove some kinds of real variation from his consideration. The reproduction schemes in *Capital* are also abstractions, but in each case the degree of abstraction is appropriate to the problem and does not obfuscate it.

To the Lysenkoist claim that genetics erects the gene as immutable and unchangeable, we reply that, on the contrary, an *essential* feature of genetics is the mutability and variation of genes. If genes were not mutable, genetics could not have been studied, for there would have been no heritable variation.

To the Lysenkoist claim that the template hypothesis of the gene assumes that God must have created the first genes, we reply, 'Remember that the conditions necessary to the initiation of some process may be destroyed by the process itself.' It is, in fact, a triumph of Soviet biology that we begin to understand the conditions for the origin of life and of pre-biotic evolution, and how the evolution of life has destroyed the possibility of present abiogenesis.

To the Lysenkoist claim that genetics erects a barrier between gene, soma and environment, we reply that, on the contrary, developmental and molecular genetics has elucidated the exact material pathway from DNA to protein to environment (the forward path of protein synthesis) and from environment to protein to DNA (the backward path of gene repression and induction), but that such pathways do not happen to include *directed* changes in DNA from environmental contingencies, because there is no material causal pathway for such directed changes, It is pure metaphysical idealism to claim that the dialectical principle of interaction demands that all possible forms of interaction must *ipso facto* exist.

To the claim that genetics does not have a 'correct' view of the internal and external conditions for change, we reply with the metaphor from *On Contradiction* that an egg will not develop into a chicken unless it is placed at the right temperature, but that a stone will never become a chicken at any temperature. That is precisely a paraphrase of the outlook of developmental genetics which asserts that a given phenotype will result only if the genes of the organism are operating in an appropriate environment, but that only some genotypes can operate with that result, no matter what the environment.

There are a number of positive contributions that a dialectical view can make in biology, but these were incompletely pursued by the Lysenkoists, or else applied at inappropriate levels.

Marxism stresses the unity of structure and process. Lysenkoists were justified in rejecting the view which sought explanation in terms of visible structures. It was valuable to expect and investigate the various physiological processes that accompanied the visible fusion of cell nuclei. But in counterposing process to structure, their view was more like that of anarchism, which sees structure as rigidity, death and enemy of process. The emphasis on process resulted in seeing the cell as a blur of interconnection among blurs. In the end, they preserved the structure–process dichotomy.

Marxism stresses the wholeness of things, both between organism and surroundings and within organism. Therefore, even among Marxist undergraduates in the 1940s in the United States, there was discussion of the need for feedback from the cytoplasm to the genes in development. But Lysenko did not seriously consider the relative autonomy of sub-systems, while genetic dogma allowed only one-way interaction. It was only much later that the modern genetic view, associated with the work of Jacques Monod, arose in which metabolites combine with some of the genes to regulate the activity of other genes. It is not clear to us whether Monod's own Marxism was relevant to the discovery.

Marxism stresses the integration of phenomena of different levels of organisation. But Lysenkoists saw only the intermediate level of the organism and its physiology. It was a one-dimensional scheme in which the intrusions of molecular events were dismissed as chance, while the level of the population or community was ignored as dynamic entities in genetics or evolution. This despite the pioneering work of Gause (*The Struggle for Existence*) in Moscow at the same time, which opened up the modern ecology of coexistence.

The view of evolution as the simple consequence of individual genetic modification meant that Lysenkoists in fact had no evolutionary theory distinct from adaptation.

Although Marxism stresses the interpenetration of an object and its surroundings, and although Lysenkoists stressed the importance of environment, they never really took it apart. They did not differentiate among regular and sporadic aspects of environment, local and widespread, short-term and long-term variations, predictable and nonpredictable aspects of environment. Therefore they could not separate the different kinds of adaptive responses at the individual and population levels.

Early Marxists had already pointed out the intimate relations of random and determinate events, in which remotely related chains of causality look like chance, random processes have determinate results, and in general the categories are not mutually exclusive. But the linking of the uncertainty principle and indeterminacy to an attack on causality and on the intelligibility of the universe led Soviet Marxists to be hostile to the creative role of random processes in evolution and therefore to be biased both against mutation as a source of evolutionary variation and against the probabilistic models of population genetics. A naive Marxism made Lysenko the enemy of change.

One way in which a Marxist viewpoint can inform scientific work is by encouraging an alternative paradigm to the analytic Cartesian method. Such an alternative stresses system properties as the *primary* objects of study as opposed to the conventional view of the main effect of separate elements, to which are added, as a secondary refinement, the interactions between them. The methodology of the analysis of variance, which separates out main effects and interactions, drives analysis in quite a different direction than does a complex systems analysis. This latter is not the same as an obscurantist holism that denies any possibility of drawing material causal connections. A major success of a complex systems analysis which derives, in part, from a conscious application of a Marxist world view, is the theory of community ecology, with its emphasis on the community matrix and on species interactions.[21]

A more common use of a Marxist approach is in the analysis of how a science has got itself into apparently irresolvable contradictions (but a Marxist analysis is not the *only* way to resolve such contradictions as the history of relativity theory shows). For example, in evolutionary genetics at the present time there are serious contradictions between observations on the genetic variation within species and all standard explanatory theories. But these theories are all equilibrium and steady-state theories which allow no role for historical processes, and they are all theories based on single genes rather than on whole genomes. When an analysis of complex genetic systems is made and when assumptions of equilibrium are relaxed, the contradictions disappear.[22]

We have described the Lysenkoist movement as a failure in several ways. By linking one's stand on scientific issues to basic political partisanship, it brought the whole repressive apparatus into genetics and had disastrous effects on Soviet biology as a whole and on many scientists individually. By depending increasingly on party and administrative support, it undercut its own anti-elitist cultural revolution potential.

It also failed to fulfil its own potential as a scientific revolution and revitaliser of agricultural technology.

The potential of the Lysenkoist movement and its failure can be traced to the same sources: the Marxist philosophical framework, which opened up exciting insights, and the administrative oppression which shut off their creative fulfilment; and behind that the social gap between rural and urban Soviet Union that produced a bifurcation of Marxism into the complex, involuted, dogmatic philosophy of the professional academic Marxists and the common-sense, naive, simplistic and often anti-intellectual folk Marxism of the Lysenkoist innovators.

The insights provided by Marxism might have been strengthened and the crudities modified if it were not for the way in which the 'two camps' model was interpreted. The confrontation between socialist science and bourgeois science was seen in the military metaphor as an implacable battle ending with victory or defeat. There was no sense of interaction. Enemy scientific writings consisted of the outrageous or of admissions. We have already pointed out how this prevented any creative assimilation of new developments in genetics. It also made partisanship the test of quality and resulted in a decline in the general level of Lysenkoist research. It established a one-way external interaction between philosophy and science in which the philosophers interpreted, blessed or condemned particular scientific views, but there was never any sense of scientific advances developing the theoretical richness of Marxism. There is some danger that the errors of the Lysenkoist movement, and the recurrent vulgarisations of Marxism that even now repeat those same errors, will inhibit Marxist scientists from making a fruitful use of their world view. We hope that a proper understanding of the history of the Lysenkoist movement will be of some help in bringing the deep insights of Marxism into the practice of science.

3

Women in Physics

Monique Couture-Cherki

For my grandmother
Marie-Vincente Menec Longuevalle,
Illiterate, who at six tended the cows
while her brothers went to school.

I feel diffident about writing this chapter: is it proper to speak of women of a given group, physicists, in isolation from the whole mass of women? Two theses have confronted the women's movements which have appeared in recent years in capitalist countries: the first, which is close to the position of the communist parties and workers' organisations, consists of making a distinction between bourgeois and proletarian women, and argues that the struggle of proletarian women should be included within the 'class struggle'. As Lenin had already put it: 'We must win over to our side the millions of toiling women in the towns and villages. Win them for our struggles, and in particular for the communist transformation of our society . . . we must train those we arouse and win and equip them for proletarian class struggle under the leadership of the Communist Party.'[1]

The other thesis consists of noting that many women can only be classed in the categories of 'bourgeois' and 'proletarian' if an equation is made between proletarian wife and proletarian, bourgeois wife and bourgeois; they are neither wage-earners nor propertied, but simply 'housewives' put in charge of the work of child-rearing and house maintenance in exchange for a reward which is a function of the social level of their husband and of his goodwill! The simple fact that I am obliged to add 'goodwill' is enough to make clear the specific feature of the situation of women. As for wage-earning women, they too generally take charge, over and above their paid labour, of the responsibility of

house and children which the organisation of our society makes a
family responsibility. This thesis goes further to analyse the situation
of women as of a caste.[2]

So far as I am concerned, while I see clearly the distinction between
Françoise Giroud (Secretary of State for the Feminine Condition) and a
woman worker in a mill or an electronics factory, I distrust the first
type of analysis, the 'class analysis', as it is in fact generally used to subordinate the women's struggle to that of the men. Society in general,
the trade unions, even more the political parties, are masculine organisations in their recruitment, their programme, their leadership;* in a
society in which only 4·4 per cent of local councillors, 1·8 per cent of
mayors, 1·4 per cent of members of parliament are women,[3] one thing
is certainly true: women are mercilessly eliminated from social life and
its decision-making bodies. And, if it is true that it is proletarian, wage-
earning women who experience the strongest oppression, each one of us
daily observes some difficulty in professional and social life and some
'traditional' attraction of the 'home'.

I therefore think that each of us, in whatever place she finds herself,
must fight to unmask those mechanisms which reject us a little . . . or
completely, and in a very specific manner, among the cares of dishes to
wash, children to look after, bread to buy, the beginning of term, the
organisation of holidays, of clothes, of homes, and so on and so forth.
And, since I am myself a physicist, I want to make some remarks on the
situation of women in the research milieu of physics to show how the
difference in their professional and social lives with respect to those of
men is linked with the role of woman as mother, and as household-
custodian, which the dominant ideology assigns them.

WOMEN AMONG THE PROFESSIONAL PHYSICISTS IN FRANCE

Women comprised, in 1974, 17 per cent of the research personnel of
the CNRS (Centre Nationale pour Récherche Scientifique) employed
within its physics sections. This percentage is comparable to that found

---

*2 women out of 17 on the national bureau of the CGT, although they rep-
resent nearly a quarter of the union members; 2 women on the national bureau of
the CFDT; no women on the political bureau of the Independent Republicans
(30 members); 7 women out of 250 members of the managing committee of the
Independent Republicans; no women on the management committee of the
Union of Democrats for the Republic; 8 women out of 81 on the management
committee of the Socialist Party; 20 women out of 123 members on the central
committee of the Communist Party (about 16 per cent) although women comprise
27 per cent of its supporters — and this last is the highest percentage.

for the proportion of women in 'higher professional ranks' in general: 18 per cent. It is clearly low by comparison with the 53 per cent of women in the population, or the 78 per cent of women in service posts, but it is also true that there are only 16·3 per cent women barristers, 13·7 per cent doctors, 2·8 per cent architects and 1·7 per cent solicitors![4] There is a rule: the more money, power and responsibility, the fewer women!

If the rule applies between the different professional categories, it is equally true within each as can be seen from the hierarchical division of women physicists within CNRS:*

Senior posts $\left\{ \begin{array}{l} \text{out of 95 research directors 8·4 per cent are women} \\ \text{out of 313 senior scientific officers 14 per cent are women} \end{array} \right.$

Junior posts $\left\{ \begin{array}{l} \text{out of 722 scientific officers 20 per cent are women} \\ \text{out of 831 research assistants 16·8 per cent are women} \end{array} \right.$

Thus it is clear that those in senior posts, holders of power (the planning of laboratories, careers of researchers, direction of research, and so on), of 'knowledge' and its associated prestige, and of the highest salaries, are largely male! It is also interesting to note the further subdivision of women within the junior posts: amongst scientific officers listed as eligible for promotion, 21 per cent are women; amongst the best qualified research assistants, 42 per cent are women.

Women's careers are thus slower and more limited than those of men; to achieve the same grades they have to be 'better' and with higher qualifications. This is also a characteristic which one finds elsewhere; the population of employed women is for the most part unqualified, but amongst employees in the middle and higher groups and the liberal professions, women are better qualified than their male counterparts (see Table 3.1).

---

*[Editors' note.] We have translated the French ranks as: senior scientific officer — *maitre de recherche*; scientific officer — *chargé de recherche*; and research assistant — *attaché de recherche*. To become a scientific officer, the French equivalent of the PhD is required; to become a senior scientific officer requires being elected from among a list of qualified individuals; to become a director, election from a list by existing directors and senior scientific officers.

TABLE 3.1
Percentage of diplomas[5]

	Among women	Among men
Employees	54	45
Middle ranks	86	74
Liberal professions and higher ranks	79	68

Concerning the division of women amongst university teachers, it has been difficult to obtain comparable figures because of the autonomy of the individual institutions, and I shall only cite the division within the UER (United Group for Teaching and Research) of physicists in the University of Paris VII, which has a supposedly liberal reputation and in which I am myself a lecturer*.

UER of physics, Paris VII, 194 teachers, 24 per cent women

Professors and readers	26; 11 per cent women
Lecturers	102; 26 per cent women
Assistant lecturers	66; 29 per cent women

Thus the senior postions are largely male, junior ones rather more female; and the male/female split between the two is a little sharper even than at CNRS.

But then, who decides on these careers? Who, in practice, makes the selection? At CNRS the commissions which decide on careers contain twenty-six members. In electronics, twenty-six men; in solid-state physics, twenty-six men; in theoretical physics, twenty-three men. All are professors, institute directors, principal scientific officers, and so forth, . . . and three women technicians! The commission on crystallography and mineralogy is the most feminine, with one technician, two engineers,† one scientific officer, and one professor. In higher education, at the national level, careers are decided by the Universities

---

*[Editors' note.] We have translated reader for *maitre de conference*, lecturer for *maitre assistant*, and assistant lecturer for *assistant*. Promotion to the lectureship grade follows entry on the appropriate list, to the grade of reader following the Ph.D. degree and entry on to two lists, one open, the other restricted; in each case the promotion is contingent on election from the list by the specialist commission of the university and the Universities Consultative Committee.

†[Editors' note.] In French, the titles of these posts ('technician' and 'engineer') are masculine. even when the holders are women.

Consultative Committee. It has three physics sections; three-quarters of its members are drawn from the senior grade, one-quarter from the junior (lecturer) grade. In the 21st section (atomic and solid-state physics) there are forty-seven men and one woman (a lecturer). In the 24th section (physical astronomy, space and geophysics) there are fifteen men and one woman (a lecturer). In the 20th section (particle, nuclear and theoretical physics) there are twenty-seven men and five women. (It appears that the world of theoretical physics is more welcoming, because of the five women, four are in the sixteen-strong sub-section on theoretical physics.) For the nomination of readers, the Federal Council of Physics, an elitist organisation, plays, in the Paris region, a very important role in classifying the candidates: its twenty-five members (twenty-two professors and three lecturers) are all men.

Among the several different universities there are local commissions of physicists which decide on the careers of teachers. At Paris VII, a reputedly 'left-wing' university, the specialist commission includes lecturers, by contrast with other universities, but there is only one woman among its thirty members! This demonstrates, in passing, that liberalism, indeed 'left-wingism', does not of itself much change the situation of women, and they are therefore committed above all to rely on their own strength.

I have spoken of the CNRS and of higher education, but I could have cited other research organisations. One thing is general: even if the proportion of women, always slight, varies from one organisation to another, none the less the managing committees are exclusively male, the fifteen members of the Administrative Council of CNES (National Centre of Space Research), the president, director-general and members of the Administrative Council of CNEXO (National Centre for Exploitation of the Oceans), the delegate-general and his three associates at the General Delegation of Scientific and Technical Research, the dozen members of the Consultative Committee for Scientific and Technical Research, which decides on research directions at the national level . . . they are all men who direct and control, relegating the women to the executive functions.

This situation becomes apparent prior to the level of the professional formation of women. The great delays in school legislation are immediately apparent: the Guizot law on primary schooling for males dates from 1833. The Duruy law, its counterpart, which obliged communities of more than 500 inhabitants to open a girls' school, dates from 1867! The normal schools for girls were opened in 1879, forty-six years after those for boys (1833); the girls' *lycées* (1880) seventy-eight years after those for boys. The science qualification (*agrégation*)* for boys dates from 1821, that of girls from 1884! Actually, women students form about a third of university classes in science (33·2 per cent in 1970 – 1), but they are also much less numerous in the prestigious institutions which form the great scientific schools in France. These are of two types: the mixed schools, which are in fact men's schools whose doors were closed to girls until recently, and the separate parallel women's and men's schools.

The *École Polytechnique*, for example, comes into the first category, it dates from the Convention (1794) and did not admit women till 1972. One of these women gained first place in the following year, and now there are about a dozen women amongst the 300 successful men.

The *École Centrale*, which dates from 1829, admitted its first woman student in 1921 and also has seven or eight women amongst the 300 successful men. The *École de Physique et Chimie*, which dates from 1887, admitted its first women in 1971, and includes about ten woman amongst its forty-five annual graduates. The *École Superieure d'Electricité* (1894) has admitted women since 1918 and, in the good years, has five amongst 300 graduates!

The *Écoles Normales Superieures* of la rue d'Ulm and of Sevres come into the second category. Ulm was founded in 1794 and is for men students; Sevres, founded in 1880, is for women. They are, in principle, of a comparable level, but in fact the maintenance of the two different schools is generally exploited in the interests of men. For example, the male graduates at rue d'Ulm are more numerous than those of women at Sevres (ten years ago the ratio was 2 to 1, the women only twice were in the majority). In actuality, the posts available in higher education and in research are only rarely available for those leaving the *École*, and it is the women who have a 'tendency' to go soon after to secondary education.

The physical and chemical laboratories at the *École*, which are relatively prestigious, are part of the men's school, and the establishment of a proper laboratory at Sevres itself, small at present, was only achieved within the last five years. The internal regulations at Sevres were for a long time stricter than those at Ulm, and until the last ten years it was forbidden to the women to receive men in their rooms;

---

*A competitive examination rather like that for the British Civil Service, for the teaching staff of all state secondary schools and some university faculties.

sexual repression was certainly made part of their training!...and there are many other aspects which could be described.

So far as the existence of two types of *agrégation*, one for men, and the other for women, one can note that in 1973 there were sixty-seven female posts in physics (physics, chemistry and applied chemistry) as against 126 for males. It is not rare that the two so-called separate but equal institutions for males and females are not in fact so 'equal'. For instance, the *École Polytechnique Feminine* is at a much lower level than the *École Polytechnique*, and recruits students with two years less training.

One can thus conclude that women's education in general, and scientific education in particular, is very far behind men's education; this is a very important point when one bears in mind the influence of diplomas on the professional activity of women. We can give as an example the rate of professional activity of women between twenty-five and twenty-nine years of age as a function of the level of their diplomas. Amongst those who do not have a diploma, 34·1 per cent have a professional activity, but amongst those with a bachelor's degree, the figure is 78 per cent, and amongst those holders of a higher diploma, 81 per cent. It is therefore clear that the level of the diploma, and of training in general, helps determine the exercise of a profession by a woman.

PROFESSION, FAMILY AND IDEOLOGY

This situation clearly shows that women physicists only penetrate an essentially masculine territory, where the classical feminine stereotypes still remain intensely powerful: the Mother, whose life is consecrated to the children, the husband and to the house — and the Sex Object, made-up, depilated and deodorised . . . photographed to sell a refrigerator or a car.*

Those scientific women who have proved themselves to be sufficiently competitive in their studies and first years in the laboratory

---

*The existence of these stereotypes is crystallised in the institution of the 'Ladies' Programme' at scientific meetings; women are thought – *a priori* – merely to accompany their conferee husbands. It is envisaged that they will visit the town centres, the infant schools, the fashion parades, and so on and so forth. In fact, scientific women are more rare at meetings than in the laboratories; they are on the whole less frequently invited than men as they 'have' (or are thought to have) more family constraints and the absence of a 'Gentlemen's Programme' often obliges them to go to the meetings alone.

cannot be sent back, even when there are children, into exclusively caring for their families; their educational standard and the salary they can obtain are the foundations of their precious economic independence and are too valuable for them to renounce. None the less a formidable pressure is exerted on them, as on all the other women in the same 'class', to maintain at the same time their professional *and* their traditional role of mother, wife and housewife.

I have demonstrated how, in the present state of affairs, jobs offered to women are few, but I must also mention the strength of the ideological barriers: first, openly declared sexism – some laboratories and some groups are 'closed' to women; second, the transformation of an objective situation of oppression into a situation of a sense of psychological guilt and inferiority permanently manufactured by the dominant ideology.

> We don't find women in top jobs? – that's because they don't know how to run things . . . . In the highest posts – generally speaking? – that's because they have no ambition. And if they are almost entirely absent from the history of physics – that's because they are not creative.

I cannot say if women are creative or not, and even if the question had a meaning in the last century, where the process of scientific production was basically still individual and in the hands of a prestigious few, it becomes a false problem at a time in which the practice of science is basically socialised, in which work has become atomised and in which the role of the laboratory chief is similar to that of the managing director.

But I also know that the history of women as beings, who completely possess a body, a sex, a desire and a creativity which are not merely a reflection of those of men, has only just begun. I also know that this creativity is eroded from the very first years by the passiveness that girls are encouraged to adopt, the discrimination between boys and girls, the 'secondary' roles that are offered to them and the absence of viable feminine models. History, when we think of it, is taught to us entirely in the masculine gender. What discretion editors and mass media use when dealing with the most striking women of our time, and in our history! Have they ever taught us at any moment about those women whose names, despite them, have come down to us today? Here I am thinking of Clara Zetkin, Alexandra Kollontai, George Sand, Lou Andreas-Salome, Flora Tristan and as many others whose very names and existence we do not know of.

And what can one say of the permanent derision and deprecation of the activities and writings of women? Leprince-Ringuet, a 'distinguished' physicist when invited to comment upon the fact that a woman passed first in the examinations at the Polytechnic, the very first year that the school opened its doors to women, declared to *Réalités*:

> To pass such an examination it is not really necessary to have a faculty of imagination, the creative spirit which plays an increasing role in modern life . . . women are intellectually docile. They much more readily accept study without opposing the contents of the course, they are much better made to prepare for difficult, competitive examinations. Hence they come top.

The General Delegate of Scientific and Technical Research told *Le Monde* in 1973 that what he liked 'about a man is his character, his competence and his humour; and about a woman . . . the freshness of her soul'.

There are, nevertheless, some women physicists who have become well known. Marie Curie*, for example, had the exceptional honour of receiving two Nobel Prizes. How is she considered? I quote again from Leprince-Ringuet, who told *Réalités* in 1972: 'Between Pierre and Marie Curie, Pierre Curie was a creator whose very genius established new laws of physics. Marie radiated other qualities: her character, her exceptional tenacity, her precision and her patience.' Public authorities are undoubtedly in agreement with him when they thought they were honouring Marie by belatedly adding her name to the street *Rue Pierre Curie* in Paris, which has for several years now been called the *Rue Pierre et Marie Curie*.

In fact, such a procedure is very widespread. As soon as a woman has an original idea, the idea is deprecated by the very fact that it was a woman who made the discovery. Should a man and woman work together and produce something interesting, the man is attributed the distinction . . . as well as the honours! † As far as 'honours' awarded to

---

*Nobel Prize for physics 1903 awarded for the discovery of radioactivity with Pierre Curie and Henri Becquerel; Nobel Prize for chemistry 1911 for the discovery of radium and the study of its properties.

†[Hilary Rose note.] A journalist once told me that he had heard Steven Rose give a 'challenging lecture' at a British Association meeting. The joke (can we call it that) was that it was me not Steven — a confusion not easily explained on visual or auditory input alone.

women and their scarcity, is it known that the only woman buried in the Pantheon* in Paris is Madame Berthelot† (who is only there because her husband requested so?).

The scarcity of famous 'women', the discretion with which they are treated and the inevitable deprecation have very important consequences for women. The system perpetuates itself by the absence of models with which women can identify. Every creative process requires that its discoverer has first conquered his identity and his autonomy. This occurs through a series of successive identifications (with parents and their substitutes, members of the family, leaders, important persons, friends, and so on) with the image of the other, which is interiorised; only after such a process is it possible to detach oneself so that one becomes a different, creative individual.

The fact that another woman from the Curie family, Irene Joliot Curie‡ has left her name in the history of physics demonstrates the importance of models! But by and large women are deprived of such models! Women physicists for the most part can only identify with men or with 'housemothers'. They find themselves lost between two types of images and cannot really identify themselves. They seek refuge in compulsively inauthentic behaviour. Those who identify themselves a bit more with men climb into the higher reaches of the hierarchy, but only by renouncing their own bodies. Those who identify themselves with the 'housemothers' retain a residual relationship to their bodies but remain in the lower reaches of the hierarchy.

Generally they too seek to realise themselves through their husbands and children, which is what the system in general conditions them to do. The consequences of this are immediate: isolated from others within the 'family', threatened with the loss of part of their identity and social position as well as of their husbands, they become apprehensive of each other and react to the aggressiveness that they experience in the patriarchal, capitalist system in its entirety, in their day-to-day relations with men, on themselves, other women and their daughters. It is the general mechanism found amongst all the oppressed, and Franz Fanon has already pointed out that the colonised, when they do not revolt against their oppressors, end up by attacking each other. Hence

---

*The Pantheon is where illustrious 'men' are buried in Paris.

†The wife of Pierre Berthelot, a nineteenth-century chemist; she made important contributions to organic chemistry and thermochemistry.

‡Nobel Prize for chemistry in 1935 with Frederic Joliot.

the rivalry between women born of the family structure in our society; the isolation of women that grows from it helps maintain the very system which oppresses them.

The only way out for women is to struggle, to overcome the last barriers and invade the social and political arena. All the constraints that govern their professional and social life have the aim of excluding them from the areas of decision and of power. Deprived of workable identifying models, blocked in their own professional lives, belittled in their creativity, made permanently inferior by the dominant ideology, made guilty by salaries that are much lower than those of their husbands, they are ripe to go 'home' and continue the washing-up, free and alone, washing millions of dishes by hand or by machine, changing millions of beds, buying tons of washing-up powder, taking children to school, getting up during the night, and so on and so forth.

We are made to believe that our 'femininity' is the barrier to our professional life, but on the contrary, our femininity is itself hindered as well as our professional life! Men make the decisions for us in the economic and social arena, and they also decide about our body!! (For example, the manifesto launched by 12,000 local councillors against the liberalisation of the law about abortion contained ten signatures of women.) It is not because we have periods, give birth, breast feed and have abortions that our professional life is blocked! It is blocked because we have accepted a sexual life governed by the power relations between men and women, the 'discreet' Tampax, repetitive childbirth, breast feeding replaced by Nestlé, illegal abortions, forbidden desires and the millions of repetitive tasks that make up 'housework', the millions of duties that govern people's lives, done exclusively and freely by women!

And whilst we have known for a very long time that 'minor domestic tasks overwhelm, stifle, brutalize and humiliate women'[6], it would be vain to hope that men, even those who lead the proletariat, will destroy this 'minor domestic economy'. It is not in their interest. Destruction and renunciation can only come from women.

4

Sciences, Women and Ideology

Liliane Stéhelin

Leave this Other to his mode of pleasure,
that might only be not to inflict ours on him.

Jacques Lacan (television 1974).

IDEOLOGY: SCIENTISM, SEXISM

Women scientists are today the target of two principal weapons in the
ideological arsenal. There is, on the one hand, an elaborate dialectic
between the ideology of the dominant class and science (and the contributors to this book have set themselves the task of dissecting this
dialectic); on the other hand, the feminine condition has universally
followed (at least in historical times and in so-called civilised cultures)
from the domination of one 'class', that of men, over another, that of
women. Now, what is ideology? I will borrow Roland Barthes' definition:

> Ideology is the idea inasmuch as it dominates: ideology can only be
> dominant . . . there is no dominated ideology; on the side of the
> dominated there is nothing, no ideology, unless — and this is the last
> degree of alienation — it is the ideology which they are compelled
> . . . to borrow from the class which dominates them. The social
> struggle cannot be reduced to a struggle between two rival ideologies;
> it is the subversion of all ideology which is at stake.[1]

If ideology is the idea inasmuch as it dominates, 'the ideology of
science' is truly the last word in ideology since there is only one idea
which can truly dominate without resort to extrinsic violence, that is
the true idea (the truth of the idea, not the idea of truth) whose privilege

science claims.[2] Sexist ideology by itself, even if it cannot argue from a basis founded on the true idea, has not on that account been less effective over the centuries, and a woman in science (women practitioners of science) seems to be situated at the crossroads of these two ideologies: on the one hand the ideology of science to which she adheres (or not) as an *individual*, on the other the sexist ideology which governs our patriarchal society; and which she confronts, as a *woman*. Courageous little soldiers, some women scientists aim to take part in the abolition of the second. But are not these crossroads a snare? Are not scientism and sexism two *forms* of the same ideology, controlled by a fundamentally masculine code? By seeing in science one of the ways of denouncing the sexist ideology, does not one succumb to an illusion, fall into a trap? Does not the ideology of science aspire to camouflage its sexist connotations by conducting a discourse which does not involve sex, a camouflage which, we will see, hardly survives a practical test?

What are these mysterious ways in which sexism and scientism diverge? The laboratory coat, the so-called prestigious uniform[3] effaces, levels, normalises the body; psychological tests of intellectual aptitude show that women, like men, possess the necessary and sufficient brains to do honest (and even sometimes very good) science,[4] even if specific personality characteristics (which it is conceded can be due to 'socialisation') handicap them;[5] yet in passing from the empirical demonstration of an absence of correlation between sex and scientific productivity,[6] the evidence of goodwill is there to proclaim, justify and demonstrate that science, in order to progress, only needs to create sexual assimilation amongst its workers.

Thus Bruno Bettelheim declared at a conference introducing the MIT Symposium on American Women in Science:[7]

> Because scientific problems and tasks are identical for men and women, since they depend, not on the sex of the worker, but on the nature of the problem, there is the pitfall that women may try to deny that they have feelings, womanly feelings, about these problems and may repress them and try to approach these tasks with the same emotional attitude that men have who for generations have been active in these fields of study and achievement. But this should not need to be so. And to disregard their specific female feelings about these tasks can be to the detriment not only of the job at hand, but also of the possibility of recruiting a large number of women for these important tasks.

The reply was: 'I wonder whether the tiny atoms and nuclei or the mathematical symbols or the DNA molecules have any preference for either masculine or feminine treatment.' Here the opposer's argument is typically ideological since he takes up Bettelheim's logical premises again as epistemological foundations, while the latter aims at going beyond them by introducing the notion of a feminine dimension in scientific practice. We are here in the presence of a phenomenon of internalisation: the internalisation of a postulate which forbids scientists of whatever sex, and under threat of seriously disturbing their internal coherence, to introduce sexual connotations into their proceedings.

Thus, if today, and over the last half century, women have infiltrated, permeated, besieged, the scientific establishment, this place where 'the last word in ideology' is spoken, what does it mean? Are we not witnessing an attempt to recapture science, a recapture complacently (unconsciously) agreed to by the establishement, which even shows a will to integrate. Why then do (some) women believe that the 'liberation' of women must take this road? Surely a really radical feminist confrontation, which aimed at subversion of sexist ideology, would involve the contrary — the subversion of the ideology of science? In other words, does the ideology of science participate in placing structural obstacles in the way of the subversion of sexist ideology? The supreme ruse of science is to support women's struggle for their liberation, while letting them interpret the obstacles they encounter as only *conjunctural*, the manifestation of the last disruptive forays of a sexist ideology which science pretends to fight. But let us suppose that 'another' science were possible, which would allow women (without excluding men) to express their identity within science, to develop their feminitude. This, unlike the present situation, would offend neither men nor women. Meanwhile, the ideological argument which must be denounced is the claim by contemporary science to sexual neutrality.

SCIENCE AND THE MASCULINE CODE

Contemporary science (space compels me to mention the historical dimension only in passing) no longer corresponds to the Socratic desire for knowledge alone. It functions as an aspect of production (production of truth),[8] itself serving the higher goal of production. It is remarkable that, transcending ideological differences, a consensus has today been reached on this pair, science/production. That a Galbraith should make a liberal diagnosis of this (un-)natural but ineluctable alliance between technostructure and science,[9] that a Richta should make the apologia for it (science, technique, production)[10] or that a Marcuse formulates a critical analysis of it, provides the evidence that is indeed an instrument which serves the code of production. 'We live and we die under the sign of rationality and of production . . . the principles of modern science were structured *a priori* in such a way that they could serve as conceptual instruments for a universe of production control.'[11]

Criticism of scientism, if it allows the *mode* of production to be brought in, rarely, however, emerges as a criticsim of the ideology of production.

A spectre haunts the imaginary revolutionary: it is the phantasm of production. Everywhere it nourishes an unrestrained romanticism of productivity. The critical notion of the *mode* of production does not touch upon the *principle* of production. All the concepts stated there only describe the genealogy, dialectical and historical, of the *content* of production, and leave intact the *form* of production.[12]

As Engels showed,[13] the establishment of the principle of production is (pre-)historically located, and I will specify further on, as much in the light of historic materialism as of psychoanalysis, why the production code appears to be fundamentally *masculine*.

If science today attempts to recruit women, it is because they represent a balance of productive forces,[14] and not because their right to participate equally with men in the progress of knowledge has suddenly been recognised. Feminine participation in the production process (inside or outside science) can only result in one of two possibilities. Women attempt to internalise the production code, an attempt leading to success or failure. In the former case, what does 'success' mean? The feminist claim which opposes the sexism present in the scientific establishment takes to the road to *assimilation*; 'liberation' is deemed to lie in the appropriation of the production code attached to the scientific establishment, forgetting that it is fundamentally masculine, and that those men and women who enthusiastically support this egalitarian movement participate implicitly in its reputation. Does not this result in those 'viraginous' women, who, sometimes, under the most exquisite feminine exterior, reveal a 'complex of masculinity', being categorised (as per Freud) as neurotics?

Alternatively there is an attempt at subversion of the production code and thus of science itself. The system insures itself against this danger by letting women (scientists as well as others) face up to what is always claimed to be the main problem of inserting women into production, namely, *maternity*.

THE SOCIAL VIEW OF WOMEN SCIENTISTS

When one surveys the literature devoted to the problem of women in scientific activity, one is struck by the descriptive, factual character of the accounts. In the most optimistic among them, there is a glorification of scientific woman, from Valentina Terechkova, first and only woman cosmonaut,[15] to Margaret Townshend, who mothers her satellites at NASA,[16] and of course the inevitable Marie Curie. All are presented as evidence of compatibility between scientific success and family success. 'They succeeded', others, nay all, shall succeed. The future is golden!

More pessimistic accounts point to discrimination in employment and in promotion (especially in the United States) and to the difficulty of reconciling scientific activity with family responsibilities. For these plaintiffs there is no need to justify the new social role which women want to play, save by a generally acceptable humanism.

The ultimate goal for women in science is not a numerical one, but a philosophical one: any person should be allowed and encouraged to pursue an education and career in whatever field he or she chooses, based solely on ability, commitment, and performance, and without regard to race, age, sex, colour, religion, politics.[17]

There is no recourse to the principle of women's participation in the development of science; analytical, theoretical research into its foundations never appears. The principle is strategically erected, and only obstacles of a tactical order, and the manner of overcoming them, are discussed.

Discrimination

This seems to operate particularly in the United States and doubtless is more dramatically resented as it manifests itself explicitly, particularly in the scale of salaries. The men themselves appear to see this as the major problem of their female colleagues, since many among them suspected Jessie Bernard's book *Academic Women*[18] to be 'a diatribe against discrimination' — actually discrimination was only one minor theme in the book — and is hard to show.[19] An abundant literature is devoted to this subject, studded with case studies and statistical facts[20] reflecting the public authorities' wish to fight it legislatively and juridically, along with other forms of 'discrimination' such as those against racial, ethnic, religious 'minorities'.[21]

Motherhood

The problems raised by maternity often appear in the same literature as that which deals with discrimination, since it is precisely the career interruptions resultant on maternity, and the organisation of time schedules which the presence of young children involves, which seem to be at the root of the discriminatory attitudes of employers of women scientists. The sanctioned solutions, always of a tactical order, consist in the socialisation of domestic work, the general institutionalisation of crèches, equal sharing of family responsibilities with the partner, and so on, all solutions intended to reduce, even to zero, the social prolongation of that anatomical characteristic which makes her, and her alone, bear the gestatory and procreative responsibility, disappear from the life of the woman-individual:

> Before this day is over, I've gotta prepare for a conference on drug biotransformation, make out an application for a 3-day conference on Programmed Instruction in the Medical Science; spend an hour driving home, cook supper, run to the cleaners, take the kid to the pediatrician, get my housekeeper's opinion on this little matter and then if time permits maybe write those Xmas cards I didn't get around to last December.[22]

And yet, and yet . . . does the solution lie *only* in these panaceas proposed to lessen the weight of household and family duties? How can one explain that, in the Soviet Union, where crèches and children's playgrounds are mainly institutionalised, women scientists have a lesser academic success and productivity than their male colleagues?[23] How can one explain that in China, despite an explicit will on the part of the government to bring women, 'half of heaven', to share the revolutionnary tasks, and to lavish on them all the help necessary to this end, 'women have not been completely liberated from their household duties'.[24] Clearly, one can set one's mind at rest and read Mao Tse

Tung's little red book where he recalls what Marx had already discerned (the emancipation of women is impossible without the emancipation of the human race), 'only in the course of socialist transformation of society will women be able to liberate themselves progressively'.

But what is one trying conceal by this recourse to pie in the sky? 'Confucius died more than two thousand years ago, but his rotten ideology, according to which men are noble and women inferior, still influences people and manifests itself all the time', is read in the Chinese press, where the campaign of criticism of Confucius and Lin Piao is still intense. But twenty-five centuries ago, in the West this time, Pythagoras also wrote that 'there is a good principle which created order, light and man, and a bad principle which created chaos, darkness and woman' and it was not by chance that Simone de Beauvoir made it one of her epigraphs in *The Second Sex*.[25] The question is: what role has woman been called upon to play from historic times, and how can she escape from it? A woman who would like to recover an original feminine identity must subvert this role and the male jurisdiction which has determined it. To want to be a woman within the framework governed by a masculine code is to be plunged into an insoluble contradiction. A woman who wishes today to lead a feminine life in the scientific establishment is facing a structural impossibility, which no organisational arrangement can resolve. If she is determined not to play the social role which history offers her any longer, if she cannot or will not internalise the masculine scientific code, she finds herself in an impasse.

WOMAN: ROLE, PSYCHOANALYSIS, HISTORICAL MATERIALISM

Woman's Social Image

Study of woman's image in different social environments reveals the 'Need for a gentle, acquiescent, motherly, devoted . . . woman'[26] What this kind of formulation disguises, and what Rocheblave-Spenle reveals in his work on male and female roles[27] is that 'in individual opinions, the masculine image, the personality traits bearing on the masculine role, appear to be centred on dominance, and the role expected of women comprises, in parallel, besides instability and lack of control, submissive behaviour . . . The stereotypes (or expected roles) are more rigid than real behaviour and they only follow the evolution of the latter slowly.' 'Only a psychoanalytic investigation would allow the intricacies of masculine and feminine roles to be unravelled'.[28] Without making such an investigation at this point, it is nevertheless necessary to recall here how psychoanalysis throws light on the foundations of the feminine role.

Woman in Psychoanalysis

For Freud, who was not at all interested in women when they were not hysterics, no woman escapes the ineluctable castration complex, which she acquires through the discovery that she is deprived of a penis. It is on the 'penis-envy' which flows from this deprivation that Freud has based his theory of feminine psychology.[29]

De Beauvoir outlines in *The Second Sex*[30] an existential criticism of psychoanalysis. I emphasis here one of the key passages, since, in filigree, one can see science involved in it:

> psychoanalysts consider that the first truth about man is his relationship with his own body and the bodies of his fellow creatures in the bosom of society; but man holds a primordial interest in the substance of the natural world which surrounds him and which he strives to discover in work, in play, in all the experiences of 'the dynamic imagination'. Man aspires to concretely reunite existence throughout the world, apprehended in all possible ways.

If science is one of the ways of answering man's existential enquiry, by knowledge of the natural world, it still remains to be known if the answer is valid for men *and* women. More particularly, has the phenomenon of alienation, that is to say the individual's tendency to search for himself or herself in objects, known objects, produced objects, an identical psychoanalytical basis in men and women? For De Beauvoir 'yes':

> primitives alienate themselves in the mana, in the totem: civilized people in the individual soul, in their I, their name, their property, their work . . . the penis is singularly appropriate for little boys to play this 'double' role . . . . Deprived of this alter ego, little girls do not alienate themselves in a tangible thing, do not compensate themselves: in that way, she is led to make her whole self an object, to pose herself like the Other. . . . If woman succeeded in asserting herself as a *person*, she would invent equivalents of the phallus. . . . Psychoanalysis could only find its truth in the historic context.

De Beauvoir thus confronts Freud and the castration complex with woman's necessity to liberate herself historically, to emerge from her condition of Other, to assert herself as a subject alienable from phallic equivalents.

Kate Millett is much more critical, making Freud the father of the sexual counter-revolution.[31] According to her, Freud deliberately rejected the social hypothesis according to which the phallus would only be envied by women for the symbolic force with which it is invested in a world dominated by men. 'Freud mistakes [purposely, according to Millett] custom for inherency, the male's domination of cultural modes for nature . . . biology and culture, anatomy and status.' Moreover, since for Freud an individual's cultural capacity is determined by the quantity of sublimated libido, and women have a reduced libido and a limited faculty for sublimation,[32] anything which women try to contribute to civilisation can only lead to nervous disorders. Still, according to Freud, women must accept the three essential elements of their personality, passivity, masochism and narcissism, corresponding to their biological destiny. Any reaction to these constitutional traits of organic nature leads to a 'masculinity complex'. What is especially interesting in Millett's criticism is that, according to her, psychoanalysis will place the scientific seal on the sexual counter-revolution, with which some cover it up:

> the influence which psychoanalysis will exercise [will be] . . . more effective even than penis-envy . . . a pseudoscientific unification of the cultural definition of masculinity and femininity with the genetic reality of male and female. Now it can be said scientifically that women are inherently subservient and males dominant.

Actually, I think Millett over-estimates the role of scientific ideology in relation to the cultural effects of psychoanalysis; it is not unanimously granted scientific status, few people really know Freud's thoughts,[33] even less his thought on women. And even when fifty years later, Jacques Lacan proclaims that '*the* woman does not exist',[34] I doubt the importance of his assertion. Freud's thoughts on woman therefore seems less interesting in respect of its sexist implications than in its crystallisation (initially for Freud himself) of the sexist ideology of several millennia, whose role is to guarantee the glorification of the phallus.

Woman in the Perspective of Historical Materialism

According to historical materialism, human institutions are not inevitable; on the contrary, they are susceptible to radical change by revolution. Engels proposed an analysis of the feminine condition in a patriarchal regime: 'The first class conflict which manifests itself historically coincides with the development of antagonism between man and woman in conjugal marriage, and the first class oppression with the oppression of the feminine sex by the masculine sex'.[35] For him, the advent of patriarchy coincides with the birth of private property, and that of the concomitant process, thanks to the instruments which man forged for himself in the bronze age, and thanks also to the discovery of slavery.

> The same cause which had assured woman her anterior authority in the home: her restriction to housework, this same cause now ensured man's pre-eminence; woman's housework disappeared from then on beside man's *productive* work: the latter was everything, the former an insignificant appendage. . . . . Therein lies the great historical defeat of the female sex.

Then there is the institution of 'monogamy [which] was the first form of family based not on natural conditions but on economic conditions, namely the victory of private property over communal, primitive and spontaneous property'. The relation between man and woman (that relation which Marx called 'the most direct, the most natural and the most necessary of a human being to a human being',[36] thus appeared to Engels as the relation of one economic class to another, the foundation of the patriarchal regime. Women's liberation thus follows for him a parallel road to that of the proletariat: women must accede to economics. If Engels' revolutionary perspective implies both the abolition of private property and of the patriarchate on which this property rests, its weakness for us lies, as argued above, in the fact that Marxist criticism stops at that of the *mode* of production (whose sexist dimension Engels has, however, clearly seen in the primary division of labour *for* production), without ever touching on the *principle* of production, which itself conceals a sexist dimension. The production code, immanent in historical materialism (as in the political economic system it criticises) appears to us to be historically borrowed by men in response to their desire for alienation from objects whose archetype is the phallus. That is why the production code can only be a masculine code, which raises a structural obstacle to the integration and the integrity of women in the cultural and economic system which it guarantees. There only remain for women the alternatives of espousing this system's code and renouncing their identity or being thrown back into their initial state of dependence, a role worn by centuries of submission by those who preceded them.

WOMEN IN SCIENCE: ADHERENCE TO THE MASCULINE CODE – SUCCESS OR DEFEAT?

At first I planned to write about all women in science: female researchers, but also female technicians, women secretaries, housewives, scientist's wives, and so on. I abandoned this status differentiation because the conclusions drawn from women researchers can constitute a kind of partial prototype for every investigation on women (potentially) employed in the contemporary production process. Why? If ideology feeds contemporary science, science, as a grateful child, brings in return substantial nutrition. Women scientists play in a certain way the role of spearhead of the feminine revolution, and it is significant that feminist diatribes rarely fail to mention the existence of the most famous among them (always Marie Curie). Hence there is an ideological virtue in exalting what some women, or a man won to the cause, calls feminine scientific successes, since these successes are brought about in the cultural broth which nourishes contemporary civilisation: if women scientists 'pass', the others will pass. This is an illusion, for nothing is in fact 'passed' apart from the exam for pupils careful to express the conventional masculine code, and the means proposed for women's liberation which the scientific establishment, as a benevolent host, pretends to encourage, is a hoax.

We have seen above that the two evils with which women scientists (like others) are burdened with are professional discrimination and the consequences (also professional) of maternity. But have not these obstacles, whose structural nature has already been brought out, an added *functional* role for the scientific establishment? One thesis of the sociology of science maintains that 80 per cent of scientists (of both sexes) only participate in routine work, feeding the 20 per cent who 'make science progress' with data[37] (this argument is contradicted by another thesis which estimates that these 80 per cent non-producers could be purely and simply eliminated without the progress of science suffering from it).[38] Is not the function of those women scientists who break their careers on the obstruction of male chauvinist institutions or the inextricable organisational problems set them by maternity, to feed these 20 per cent? And should one not ask if these 80 per cent do not have a second and third function? On the one hand to serve as a receptacle of knowledge elaborated by the light infantry of the 20 per cent, a receptacle for the diffusions via teaching (teaching in the broad sense, comprising that of children in the family framework, which involves the 'role' given to women); on the other hand that of being an instrument for exalting the elites, and through the elite, science itself, the challenge which the latter offers to the laws of nature, the ethic of knowledge.[39]

The golden doors of the scientific establishment are open to women, but they are only the doors of its antechamber. Only a few will pass beyond the next doors, to the great astonishment of some: 'any performance by a woman is considered unique'.[40] At what price will they pass through them? Why are they so few? Let us return to the thesis that every mode of production is of masculine origin. If Marx applied himself to the denunciation of the fetishism of goods and of money, on which the capitalist mode of production rests, today 'this fetishism has become the one and constant object of contemporary analysis .... Apparently only psychoanalysis has escaped from this vicious circle by re-attaching the fetishism to a perverse structure ... of denying the difference between the sexes.[41]

It is thus necessary, in order to avoid ideological subversion, that the production code should raise this negation of the differentiation of the sexes to a postulate; but a negation defined with reference to the characteristics of those who themselves established this production code: men. Nowadays, in these feminist times, it will be said that women are the equal of men; a syntactic reversal will seldom be made which will make individual men equal to women! What is the basis for this perverse structure: to produce in order to escape the fear of castration, to prevent the Other (women) from asserting herself as a person, to forget, by mastering Nature, that it is a woman's belly that gives life (the alternative term being Oedipus, love of the producer body)? We will leave to pscychoanalysis and its various schools the duty of trying to reply. Fetishism, production, science, are all linked according to a phallic code — 'the no-penis is no-knowledge'.[42]

The (supposed) saturation of productive forces obliges women nowadays to participate in production. In private enterprise it is going to be easy, in the name of economic argument, to keep women in odd jobs (child-births are patches in the life of a woman worker which, it will be claimed, prevents running the risk of giving them highly specialised posts). In the scientific establishment the economic argument is more difficult to advance, the more so since research is often taken over by

the state. Women will therefore infiltrate there more easily, up to hierarchic levels equivalent, or at least close to, those reached by men.
Flattened, levelled in their feminine identity, become scientists like the
'others' (men), they suddenly see their career progress in crisis. The
absence of this crisis would be the non-confrontation with maternity,
which will usually signify their definitive adoption of the masculine
code. Note that unmarried women scientists greatly outnumber their
masculine counterparts (40 per cent versus 12 per cent; the same for
separated and divorced — 15 per cent versus 4 per cent).[43] Eight years
after their doctorate 50 per cent of American women scientists studied
by Astin in 1969 were unmarried while only 6 per cent of the American
female population of the same age were in this state.[44] It could be
argued that marriage is not a prerequisite of maternity, but, still in the
same study, women scientists have few children in relation to the
average for women, and women who have no children are twice as
numerous among the scientists; they escape from the historically
defined role, but their attitude is reactive — they adopt the code of
those who imposed this role, and flee from their identity.

The crisis comes at the time of the first child-birth(s). The woman
scientist then finds herself faced by two obstacles, one of an ideological
nature, the other which we will qualify as structural. To eliminate these
obstacles would signify a major defeat for the masculine code.

The ideological obstacles consists of maternity placing a scientist
under the necessity of accepting or denying the role expected of a
woman, the role of the woman-educator, woman at the hearth, submissive woman. There is then a conflict between the role of the scientist,
and the role which millennial sexist society imposes on her.

The structural obstacle consists of the woman scientist, having pro-
created, needing to continue nevertheless to play a game according to
masculine rules. Some succeed. So much the better for them perhaps,
for science in any case, and thus for the ideology. And more besides; on
the one hand because they will constitute an increased work-force,
alienated from the scientific establishment, and on the other hand
because they will be the best guarantee, in the education that they will
lavish on their children (and especially their daughters), of the perpetuation of the system. By making science the instrument of their
'liberation', and in seeing the obstacles they encounter (discrimination,
maternity, and so on) as the last resistance of a sexist ideology whose
subversion will henceforth be undertaken, are they not deceiving
themselves and unwittingly drawing 'half the men' (cf. a women's

movement slogan: 'half the men are women') to their definitive alienation and the loss of their feminine identity?

Most do not succeed. Why? Following Freud, can it be said that they
find in maternity compensation for their 'penis envy', until now sublimated in the taste for knowledge. Or rather that maternity enables
them to discern that there is another route than science, than production, to knowledge. If that is so, giving them the possibility of self-
expression in this other way would put the scientific establishment and
its functional code in serious jeopardy. The scientific institution and
the production system which it supports protect themselves then by
the game of a sort of liberal economy which leaves women confronted
by what are at first sight conjectural obstacles (child-minding, housework, homecrafts). The only ones who will be able to surmount them
will have espoused the structure of the system.

Is there another way? Another way where science would no longer
signify the instrument of man's *power* over humanity, no longer signify
knowledge constituted in the service of *production*. Scientists participate in the front rank in perpetuating a civilisation locked into the
triangle of power—production–phallus. Women scientists possibly
represent one of the key elements in the subversion of the system. The
fact that so many women scientists 'fail' permits the raising of the
Utopian veil from this subversive question; in their discouragement
before the obstacles they encounter, in their lack of ambition and
disinterest in their work, one must clearly recognise the expression of
the dominance of sexist ideology, and the mark of the internalisation of
this ideology by its own victims; but one must also see there the indication that the *active* adoption of the masculine code (and no longer only
submission to it) in social practice is intrinsically difficult for woman, if
not, for some of them, impossible. It is these women who are the
promise that one day other women (with other men?) will be able to
open the way for a new science.

5

History and Human Values: A Chinese Perspective for World Science and Technology

Joseph Needham

The first conception of this chapter was that it should discuss the relationships between human values and science in the contemporary world, especially in China. But the more I thought about it, the more I came to feel that we needed not 'a perspective for Chinese science' but a 'Chinese perspective for world science'. It seemed to me that certain Chinese values might be vitally helpful to humankind face to face with its own embarrassing knowledge.[1]

First, then, we might consider what really have been the relations between Chinese culture, and science, technology and medicine. What has the work of the past decades done, since the Second World War, to elucidate their relationship? And, third, most important of all, what clues can we get about the help which the Chinese tradition and contemporary China could perhaps give to the *ethos* and operation of the World Co-operative Commonwealth[2] of the future?

At the outset, of course, I have to declare, as is done in the House of Commons, a private interest; yet at the same time a certain authorisation for tackling these subjects. I am, then, a scientist by profession and training, a biochemist and embryologist, whose life nevertheless, at sundry times and places, came by chance into close contact with the fields both of engineering and medicine. For four years during the Second World War I was Scientific Counsellor at the British Embassy in Chungking, directing the Sino–British Science Co-operation Office (Chung-Ying Kho-Hsüeh Ho-Tso Kuan), that is, a liaison bureau which kept the scientists, engineers and doctors in beleaguered China in close touch with the free world of the Western Allies.[3] I had been 'converted' (if the expression is permissible) to an understanding of the Chinese world outlook by friends who had come to work some years before in my own and neighbouring Cambridge laboratories for their doctorates, especially Lu Gwei-Djen, today my chief collaborator. This was the origin of the *Science and Civilisation in China* project,[4] which has taken me away from experimental researches for some thirty years past. Eight volumes have appeared, and the ninth is now going through the press, so one can calculate that my collaborators and I have succeeded in producing one every 3·25 years. Since seven or eight more volumes are in active preparation, I have little difficulty in computing the probabilities of my being able to correct the proofs of the last volume *in propria persona*.

Since the series began to come out, two very great changes have taken place. When I first decided to become a full-time historian of science, I had not the slightest idea how topical the subject of Chinese science would turn out to be. A new 'Chinoiserie' period has almost come about. People simply do not know enough about Chinese culture, yet everywhere they are avid for information. We never thought that we of all people would be called upon to supply it. Our original question was: why had modern science originated only in Western Europe soon after the Renaissance? *Mais faites attention; un train peut cacher un autre*! We soon came to realise that there was an even more intriguing question behind that, namely why had China been more successful than Europe in gaining scientific knowledge and applying it for human benefit for fourteen previous centuries?

We wanted to know many things. For example, how far did the Chinese get in the various particular sciences before the era of oecumenical world science began; and again, what, if anything, did they contribute to the origins of modern science itself? these are not questions I can answer in a few words here. All I want to say is that while we were sapping and mining to discover vast stores of information which had never been appreciated by the world as a whole, because still buried within the bosom of the ideographic language, historical events occurred in China leading indubitably to her elevation to great-power status. Hence the feeling shared so widely all over the world that we must know more about Chinese culture and the Chinese.

I would like to return in a moment to these historical aspects, but before that I want to mention the second great change that has come about during the past thirty years – I mean that powerful movement away from science and all its works which characterises what has been

called the 'counter-culture', and which is widely prevalent among the
youth of the Western world at the present time. It is found not only in
the West but also to some extent in the underdeveloped parts of the
world, and I should be inclined to call it a deep psychological aversion
from 'big technology' and the science which has given rise to it.[5] Moreover, the 'disenchantment' with science cuts across all political boundaries, because the inhumaneness of science-based technology is felt to
a certain degree in the socialist countries as well as the capitalist ones.
This is something on which there is much more to be said, but meanwhile my collaborators and I would like to declare for ourselves that we
have in no way lost faith in science as a component of the highest
civilisation, and we believe that it has done incalculably more good than
harm to human beings. Indeed its development from the first discovery
of fire onwards has been a single epic story involving the whole of mankind, and by no means separated out into incommensurable Spenglerian
cultural entities.[6] At the same time it is only too clear that modern
science and technology, whether in the physical, chemical or biological
realms, are now daily making discoveries of enormous potential danger
to humans and their society. Their control must be essentially ethical
and political, and I shall suggest that this perhaps is where the special
cultural tradition of the Chinese people could affect the entire human
world.

THE HISTORICAL BACKGROUND OF SCIENCE IN CHINA

May I now return for a short time to the historical development of
science and technology in China. I think it would be fair to say that the
history of science, technology and medicine in Chinese culture has now
really obtained *droit de cité*; it has become accepted and established (I
hope not in any bad sense). It is worth looking briefly at a few details
of the Chinese achievements. The mechanical clock has by long tradition been accepted as one of the most outstanding achievements of
European mechanical genius at the close of the Middle Ages, and something without which modern science would never have been possible —
but it remained to be shown that some six centuries of hydro-mechanical
clockwork had anticipated the first clocks of Europe. Indeed one can
say that the soul of the mechanical clock, the escapement, was not the
invention of an unknown artisan about ·+1280 in Europe, but rather
that of a Tantric monk and mathematician, I-Hsing, and his collaborator
the civil official Liang Ling-Tsan, in China in +725. Here there was a

close connection with mechanical models of the heavens continuously rotated, a development which arose as early as the +2nd century in China as a result of the fact that astronomy there was polar and equatorial rather than ecliptic.[7]

For nearly two thousand years before the beginning of modern science there were no people in the world who observed the changes in the heavens more persistently, and recorded them with greater amplitude and accuracy, than the Chinese.[8] I read recently an essay by the physicist Victor Weisskopf in which he spoke about the termination of the evolution of certain stars in those tremendous explosions which are called supernovae.

> One of these [he said] occurred in the year +1054 and left behind the famous Crab Nebula, in which we see the expanding remnants of the explosion with a pulsar in the centre. The explosion must have been a very conspicuous phenomenon in its first days, surpassing the planet Venus in brightness. So different from today's attitudes was the mental horizon in Europe at that time that nobody found this phenomenon worth recording. No records whatever are found in contemporary European chronicles, whereas the Chinese have left us meticulous quantitative descriptions of the apparition and its steady decline. What a telling demonstration of the tremendous change in European thinking that took place at the Renaissance![9]

This is perfectly true, and radio-astronomers today habitually make use of the Chinese records, at least as far back as the beginning of our era. Many other examples could be taken from these branches of science. For one thing only, the complete description of the phenomena of parhelia (mock suns and Lowitz arcs) was given in China in the +7th century, but not in Europe until the +17th.[10]

Again, turning to technology, one could look at the morphology of the reciprocating steam-engine. It is, I am sure, not widely appreciated that this was in all respects completed in China some 500 years before the time of Newcomen and Watt.[11] The great difference was that the direction of energy transfer by interconversion of longitudinal and rotary motion was the opposite to the steam-engine, since the old Chinese machine was driven by its water-wheel while its piston-rod worked upon metallurgical furnace bellows. Conversely, in the steam-engine, first steam and vacuum, then steam alone, acted on the piston-rod, which accordingly, by an identical conversion, produced useful rotary motion. In China one can find a step-by-step evolution of all the components.[12] The eccentric came first of all, in the −4th century, for the rotation of querns; the crank-handle in the +2nd century; the eccentric and connecting-rod, so arranged that several people could exert power on the quern's rotation at one time, in the +5th century and then the addition of the piston-rod in the +6th century allowing the application of water power to bolting machines for cereal sifting.[13] By the +13th century, and probably a good deal earlier, the assembly was employed for the blowing of furnace-bellows, a fact which may well be one of the reasons for the astonishing development of the iron and steel industry in the Sung period.[14]

I must say that the achievements of traditional China in science, technology and medicine continue to surprise even ourselves all the time. Just recently, for example, we have resumed our work on the medical volume and have been drafting the sections on acupuncture and other medical techniques. In the course of this we have naturally had to say a good deal about the ancient Chinese theories of the circulation of the *chhi* (*penuma*) and the blood in human beings. I myself was at one time inclined to believe that those interpreters were right who had seen the ancient Chinese circulation as being a very slow one of 24-hours duration, but much closer study of the essential passages in the *Huang Ti Nei Ching* convinced us that as far back as the −2nd century the Chinese physicians had a conception of the circulation of the blood only sixty times slower than the Harveian circulation which we understand today. Many have long appreciated that Chinese civilisation was far more circulation-minded than any of the other ancient cultures, such as that of the Greeks,[15] but I was not prepared to find them so close on the heels of modern physiology.

Second, we have been deeply impressed with the brilliance of the tie-up between the acupuncture tracts and points on the surface of the body and the state of affairs, pathological or otherwise, in the viscera and other internal organs. A full two millennia before modern physiology, the ancient Chinese physicians were aware of phenomena which today we would describe as viscero-cutaneous reflexes, referred pain, and those areas of nerve distribution on the surface of the body which we call Head zones and dermatomes.[16] Since the ancient and mediaeval Chinese scientific men, engineers and doctors have been consistently surprising us ever since we started on this work at the end of the Second Word War, I must say I am now expecting them to continue to do so for the rest of my life — and for very much longer if Westerners are still willing to look into the work of their Chinese predecessors.

I hope truly that they will be large-minded enough to do this because, after all, as the Romans said, humanity is one, and nothing human should be regarded as alien from us. I dare say it will take a certain high-mindedness, because I have had many opportunities of observing that our work has been irritating for conventional Western minds; the achievements of science and technology were often what they were most proud of racially, and to show that others had done as much or more was hurtful to pride.[17] But pride is a thing we can always do without. It weighs little in the balance of Anubis compared with love and friendship.

Such was the grudging attitude of the establishment, of course, in the West, but now, as time has gone on, we are faced with an entirely different and opposite reaction, that of the younger generation and the counter-culture, who would be inclined to say that if Chinese civilisation had done so much for science and technology, that would give it a bad mark in their estimation. So you cannot win. But that does not bother us, because we believe in science (though not in science alone) and we hold fast to what we regard as true historical perspective, without which no understanding of the past is possible.

In the light of all this, some estimates of China with relation to science need complete revision, in fact relegation to the scrap-heap. The idea, for example, that Chinese scientific development has been 'just a transplant', something out of keeping with the cultural inheritance, is sheer nonsense. The idea that China (or Japan, or Africa, or anywhere else for that matter) will be necessitated to adopt the ways of life of the Western capitalist world as a whole in adopting science and technology is also nonsense, whatever may have been the superficial situation so far. Look at the almost total failure of the Church as a human institution in Asia. I found when I was in China during the war that people often used to refer to modern science as Western science (*Hsi-yang kho-hsüeh*), and I always used to say in speeches that on the contrary it should be regarded as *Hsien-tai ti kho-hsüeh*, distinctively modern science, open for the participation of all trained men and women, totally irrespective of race, class, colour, creed, and so on.[18] Long afterwards I found that this attitude had been exactly the same as that adopted by the Khang-Hsi emperor in +1669, when he had insisted that the title of the work on mathematics and calendrical science prepared by the Jesuits should be *Hsin Fa Suan Shu*, and not *Hsi-Yang Hsin Fa Li Shu*, as it had been twenty years earlier.[19] Though the emperor probably knew very little of the Royal Society, founded just at this time, he recognised with great acumen that the 'new, or experimental, science' was essentially new and not essentially Western.

Of course the rate of growth in science and technology differed greatly between China and the Western world. The Chinese rise was slow and steady, it never had anything corresponding to the 'Dark Ages' which Europe went through.[20] Some of the Greek contributions (perhaps Aristotle and Euclid) rose above the Chinese level, but between the +4th and +14th centuries the level in Europe was abysmally below it; only with the Renaissance and the unleasing of modern science did the European curve rise quickly above the Chinese one and pursue its exponential increase to bring about the world we know. There was thus what might be called a *décalage* in the development; but there is one striking thing to note, namely that some of the greatest Chinese inventions took place precisely during the period when Europe was at its lowest ebb. For example, block-printing was going by the +8th century and moveable-type printing by the +11th. The formula for the first known chemical explosive, gunpowder, was established in the +9th century; and the magnetic compass, which had existed long before as a geomantic instrument, was adapted to navigation in the +10th. These were the three inventions singled out by Francis Bacon.[21] Clockwork too, as I have already mentioned, took its rise in the +8th century, about the time when scholars like Isidore of Seville and Beatus Libaniensis were trying to save what remained of ancient knowledge, and when the number of plant species known — quite a good index of scientific capabilities — sank to its absolute minimum in Europe.[22]

THE COUNTER-CULTURE AND THE ANTI-SCIENCE MOVEMENT

Let us now return to the anti-science movement and the counter-culture.[23] Some of the most telling formulations of the disenchantment of the young with science have been presented by Theodore Roszak in his books, *The Making of a Counter-Culture*,[24] and *Where the Wasteland Ends*.[25] He and the young are against modern science because they feel that it has had evil, totalitarian and inhuman social consequences. They are not content to put this down merely to misapplied technology; their criticism of science itself goes deep. They attack 'the myth of objective consciousness,[26] detesting that 'alienative dichotomy' which separates the observing self from the phenomena in Nature,[27] and sets up what they call an 'invidious hierarchy' which raises the observer to an inquisitorial level, free to torment Nature, living or dead, in whatever way will bring intellectual light.[28] They feel, too, that science encourages a 'mechanistic imperative', that is to say an urge to apply every piece of knowledge, in every possible way, whether or not its application is health-giving for human beings, or preservative of the non-human world in which they have to live.[29] The scientific world-view is thus accused of a cerebral egocentric mode of consciousness, completely heartless in its activity. It is not as if scientific methods of control were applied only to non-human nature; the 'scientisation of culture' is calculated to enslave humanity.[30] There are many techniques of human control, such as the behavioural and management sciences, systems analysis, control of information, administration of personnel, market and motivational research, and the mathematisation of human persons and human society.[31] In a word, technocracy is rampant, and the more complete the domination of Nature, the more fully does it become possible for ruling élites to increase their control of individual human behaviour.[32]

Since Francis Bacon's time the essence of the scientific method has been alienation, in the sense of an absolute distinction between the observer and the external world, with which he can have 'no sense of fellowship nor personal intimacy, nor any strong belonging'.[33] Nothing inhibits the ability to understand, and, after understanding, to manipulate and exploit to the full – hence many of the cruelties which have undeniably been perpetrated in the name of modern biology and physiology. As I write these words there is news of a demonstration at the annual meeting of a great British chemical industry, protesting against the use of dogs as experimental animals in studies of the cancer-producing properties of tobacco.[34] There is much that can be said against the 'callousness' of science, yet without what one might call 'clinical detachment' no scientific medicine would ever have come into being. In the same way the opponents of science cannot deny its pragmatic value in that pharmacological knowledge does lead to the relief, or cure, of disease, and that flight would be impossible without a knowledge of aerodynamics and thermodynamics. The anti-science movement is in rather a quandary here for it can hardly wish mankind to return to the infancy of pre-scientific ignorance, while at the same time it is justifiably uneasy, indeed outraged, at the uses which are constantly made of scientific knowledge;[35] and filled with fears for the future.[36]

THE FORMS OF HUMAN EXPERIENCE

I am inclined to think that the real meaning behind the anti-science movement is the conviction that science should not be taken as the only valid form of human experience. Actually, philosophers have been calling this in question for many years past, and the forms of human experience – religion, aesthetics, history and philosophy, as well as science – have been delineated in many integrated surveys.[37] Roszak himself hints at this, as when he denies that scientific objectivity can be 'the only authentic source of truth', or when he says that 'we must be prepared to see the truth as a multi-dimensional experience.[38]

Some of his most interesting insights arise from an exegesis of William Blake, where he demonstrates that Blake's hatred of what he called 'single-vision' was really a criticism of the estranged monocular scientific experience considered as the only possible manner of apprehending the universe.[39] Many other expositors of Blake have shown that his image of the 'Four Zoas' was really a recognition of the diversity of the forms of experience.[40] For example, he wrote,

> Four mighty ones are in every man; a perfect unity
> Cannot exist but from the universal brotherhood of Eden,
> The Universal Man.

In this peculiar psychology of mythical entities, Urizen stands for the cold scientific reason, Luvah for energy, passion, feeling and love, Urthona for prophecy (aesthetic?) and Tharmas for the power of the spirit (history and philosophy?). Some of these have consorts, *shaktis* or *alter egos*. Thus, most penetratingly, Blake named the 'fallen state' of Urizen as Satan, the god of brute unfeeling power, the spirit of the dark Satanic mills with their 'cogs tyrannic'. So also the fallen state of Luvah was Orc, the chained beast of thwarted desire, religious persecution and repressive or revolutionary terror. Urthona contains in itself (hermaphrodite) a Yin/Yang pair, Los and Enitharmon, and all these war with one another, or become reconciled with one another, on the battlefield of Albion, the individual man.

Among the most interesting recent treatments of the diverse forms of human experience is that of Victor Weisskopf, who suggests that they stand to one another in a relationship something like the Heisenberg uncertainty principle in physics, the dilemma of indeterminancy or complementarity.[41] If one tries to investigate the details of the quantum state by some sharp instrument of observation, one can only do so by pouring much energy into it, and so destroying the quantum state. The necessary coarseness of our means of observation makes 'exact' observation in the old sense impossible. One can either know the speed or the position of a sub-atomic particle, but never both at the same time. Weisskopf writes:[42]

> The claim of the 'completeness' of science is that every experience is potentially amenable to scientific analysis and understanding. Of course many experiences, in particular in the social and psychological realm, are far from being understood today by science, but it is claimed that there is no limit in principle to such scientific insights. I believe that both the defenders and the attackers of this view could be correct, because we are facing here a typical 'complementary' situation. A system of description can be complete in the sense that there is no experience that doesn't have a logical place in it, but it could still leave out important aspects which in principle have no place within its system. . . . Classical physics is 'complete' in the sense that it could never be proved false within its own framework of concepts, but it does not encompass the all-important quantum effects. There is thus a difference between 'complete' and 'all-encompassing'.
>
> The well-known claim of science for universal validity of its insights may also have its complementary aspects. There is a scientific way to understand every phenomenon, but this does not exclude the existence of human experience that remains outside science. . . . Such complementary aspects are found in every human situation.

Weisskopf goes on to say that whenever in the history of human thought one way of thinking has developed with force, other ways of thinking or apprehending have become unduly neglected, and subjected to an overriding philosophy claiming to encompass all human experience. This was obviously the case with religion and theology in mediaeval Europe, and it is certainly the case with natural science today. And he goes on to draw the very existentialist conclusion that

> the nature of most human problems is such that universally valid answers do not exist, because there is more than one aspect to each of them. In both these two examples great creative forces were released and great human suffering resulted from abuses, exaggerations and neglect of complementary ways of thinking.[43]

Indeed this is most true, and the only way forward is the existentialist realisation that the forms of experience, which have a habit of contradicting each other flat, are all basically inadequate ways of apprehending reality, and can only be synthesised within the individual life as lived. Here mutual control and mutual balance are of the essence. Could not we hope to learn something of this from our Chinese friends? As Weisskopf puts it: 'human existence depends upon compassion, and curiosity leading to knowledge; but curiosity-and-knowledge without compassion is inhuman, and compassion without curiosity-and-knowledge is ineffectual.[44] This is a *philosophia perennis*. We find it again in Julian of Norwich, writing soon after +1373: 'By reason alone we cannot advance, but only if there is also insight and love.'[45]

The whole anti-science movement has arisen because of two characteristics of our Western civilisation: on the one hand the conviction that the scientific method is the only valid way of understanding and apprehending the universe; and on the other hand the belief that it is quite proper for the results of this science to be applied in a rapacious technology often at the service of private capitalist profit. The first of these convictions is held as a semi-conscious assumption by a great many working scientists, though formulated clearly only by a small number;[46] at the same time it spreads widely through the population, often leading to great callousness and insensitivity in personal relationships, quite beyond the power of the traditional codes of religion and ethics to modify. Similarly, the mass-production technology of the capitalist world, so freely paralleled and imitated in the Soviet Union and the Socialist Republics of Eastern Europe, has indeed supplied the peoples of the developed world with a vast wealth of material goods,[47] but only at the cost of debauching their aspirations, limiting their freedoms, and imposing controls every day more insidious and unhealthy.

CHINA, EUROPE AND 'SCIENTISM'

This brings me to the first of the fundamental points which I want to make. The contribution of China to the world may be particularly valuable because Chinese culture has never had this post-Renaissance 'scientism'. Since modern science did not develop in China orginally, but only in Europe, there was never any necessity for a Chinese William Blake to oppose the 'single-vision' of Isaac Newton's conception of the universe, nor did a Chinese Drummond or Kropotkin have to do what he could to oppose a Chinese Huxley. In spite of the debate about scientism in China in the first decades of this century,[48] Chinese culture has never been really tempted to regard the natural sciences as the sole vehicle of human understanding. This was all the more the case because for twenty centuries an ethical system not based on super-natural sanctions had been completely dominant in Chinese society; and it was history, not theology and not physics, which had been the *regina scientiarum*. This dominance of morality is perfectly represented in China in the present day by the slogan 'put politics in command', for this essentially means human moral values, and that in practice means the health and well-being of your brother and sister at the bench, in the field, on the shop-floor, and next to you in the office or at the council table. In China you will not be tempted to regard him or her as 'nothing more than' a behaviourist automaton, or a walking flask of amino-acids and enzymes, because, by the wider wisdom of Chinese organic huma-nism, you know that he or she is a lot of other things as well, for the full reality of which mechanistic or reductionist scientific explanations, however successful for limited ends, can never suffice.[49] This is my first fundamental point. It may be that scientism, the idea that scientific truth alone gives understanding of the world, is nothing but a Euro-American disease, and that the great contribution of China may be to save us from the body of this death by restoring humanistic values based on all the forms of human experience.

It may be that the roots of the matter, why Europe has suffered so much from this, go back very far. Roszak, in the second of his books, has traced the desacralisation of Nature,[50] which has been so important for modern science, to the ancient Jewish anti-idolatry complex inherited by Christianity.[51] This led, he suggests, to what one might call 'nothing-but-ism', that is (in its first form) the conviction that the statue of a Greek or Roman god was nothing but a statue, totally devoid of all magical or worshipful properties. Islam, of course, inheri-ted this too; 'Thou shalt have no other god but God, and Muhammad is his prophet.' And Christendom and Islam both inherited the aggressive intolerance which Hebrew monotheism had from the beginning infused into this issue. But to modern minds, especially those acquainted with Buddhism or Taoism, it would seem that far too much fuss was made about idolatry by the Christian theologians, rising to a crescendo with the Protestant Reformation (so significant always for the birth of modern science), because in a way it never really existed at all — for every intelligent person the statues must always have been thought of as symbols and temporary residences of the god or spirit invoked. Even quite simple people must have understood this. Certainly it was the attitude of the Neo-Platonists.[52] But the point is that this theological 'nothing-but-ism' was a prefiguration of the reductionism of modern science. It was a mental predisposition, a *praeparatio evangelica scientifica damnosa*.

First, for example, there was the suppression of secondary qualities from Galileo onwards, then, after the coming of modern chemistry, there was the belief that all the phenomena of life and mind could be explained without residue by the properties of atoms and molecules, ultimately of the sub-atomic particles themselves. We are not complaining that all this banned magic entirely from the world, or denied to Nature its authentic enchantment, because that is just the sort of thing which science has to do — within its own realm — but we are saying that the anti-idolatry complex paved the way through the centuries for the 'nothing-but-ness' of mechanical materialism and scientism. It attacked all the lesser sacralities in the name of the god of Abraham and Jacob, and this was all right as long as the creator diety of the Hebrews remained secure on his celestial throne, but when belief in the supernatural collapsed under the attacks of mechanical materialism, nothing holy was left anywhere.

How different was the situation in East Asia. Since the Chinese were never among the 'People of the Book',[53] like Jews, Christians and Muslims, they never had the anti-idolatry obsession. The holy was never identified with the supernatural because in Chinese thought there never was anything outside Nature. The numinous, the holy, could be, and was, present at many points within and beyond the world of man. 'The world is a holy vessel', says the *Tao Tê Ching*, 'let him that would tamper with it, beware'.[54] Thus, once again, Chinese culture was not under the same drives as that of Europe, and wherever you look, whether in Chinese organic philosophy, or the beauty of Nature as interpreted by the Chinese artists and poets, or in the studies of Chinese historians, this 'nothing-but-ism' never came up — and most important of all it never dominated in Chinese science and technology, great and world-shaking though their discoveries and inventions were. It may well be that this absence was one of the limitations, like the presence of that tension between the separated sacred and secular in Europe, which prevented the rise of modern experimental science in China, and encouraged it in the West; but that is no ground for doubting that Chinese level-headedness may now be deeply needed and called upon to rescue the Western world from the slough of mechanical materialism and scientism which it has fallen into.

If the anti-idolatry complex of Israel and Christendom had been the only influence leading to reductionism in Europe, the compulsion might not have been so strong, but it was reinforced from quite a different quarter, the atomism of the Greeks. The schools of Democritus and Epicurus held, broadly speaking, to a mechanical materialism, and it was put into prophetic imperishable poetry by the great Lucretius. One is accustomed to think of a schizophrenia of Europe, torn between the theological world-picture and that of materialist atomism, but in this particular they were closely allied; for while the former reduced the holy objects of antiquity to common wood and stone, the latter reduced not only wood and stone but also flesh and blood to the fortuitous clashing of hard impenetrable massy particles. Again in China what a difference! Buddhist philosophers must have talked much about the advanced atomic speculations of India, but never with any perceptible effect, for the Chinese remained perennially faithful to their prototypic wave-theory of Yin and Yang, universal transmission in a continuous medium, and action at a distance. The Yin and Yang, and the Five Elements, never lent themselves to reductionism because they were always inextricably together in the continuum, and in the organic phantasms which came and went within it; never separated out, isolated or 'purified', even in theory. So once again there was no 'nothing-but-ism' in Chinese thought to parallel what Greeks and Hebrews joined together to create in the West.

Nothing that I am saying here, however, should be taken to indicate that there is anything different about the methods of modern science in China today from those in the rest of the world. Recently, when lecturing in Brussels and Ghent, I came across a curious idea that the nature of modern science in China today was as different from that in the rest of the world as modern science had been, in its turn, from ancient or mediaeval science. This was a notion doubtless stimulated by the success achieved in China in drawing peasant-farmers and working people of all kinds into the field of observation and experiment, but it has no force. Science is one and indivisible. The differences are essentially sociological — what you do science for, whether for the benefit of the people as a whole, or for the private profit of great industrial enterprises, or for the development of fiendish forms of modern warfare; in a word, your motive. The differences will also be great according to whom you get to do it, whether you confine it to highly trained professionals, or whether you can use a mass of people with only minimal training; and this means again how you manage the whole affair.

E

But, basically, modern science is surely identical under every meridian. There is only one logic of controlled experimentation, only one application of mathematical hypotheses, and their testing by statistical methods. There are canons here which cannot be transgressed; the basic method of discovery itself, which was discovered in Galileo's time. The 'paradigms' of course change, and have changed, as knowledge increases, for example when the Einsteinian world-system was superimposed on the Newtonian one, but that does not alter the basic character of the scientific method itself. Or again, beliefs that were held in ancient and mediaeval times can come obliquely true in the light of later discoveries; for example, on the atomic level, the dream of the alchemists of transmuting baser metals into gold was bound always to remain a dream, but after the discovery of radioactivity it became possible to turn one element into another, including gold, by the addition or subtraction of nucleons. No, science is a unity, and it is not different in China from anywhere else, but what we can rightly object to is the idea that science is the only valid way of apprehending the universe. Perhaps we fell into this mistake because modern science originated among us in the West; conversely the Chinese never had the temptation under which we fell, and now is the time for them to give us help to climb back to the realm of true humanity.

THE CHALLENGE OF TECHNOLOGY AND MEDICINE

Let us turn now to an entirely different aspect of the situation. Even in a properly balanced human society, where the natural sciences were counter-balanced all the time by what used to be called in Cambridge the 'moral sciences', and other forms of human experience such as the religious and the aesthetic, there would still be great difficulty in dealing with the all but intolerable ethical choices which applied science places before mankind, and will increasingly place as time goes on. The young people of the counter-culture are revolted by the necessity of making such choices, but neither they nor we can go back to the 'bliss of ignorance' as in primitive times. Actually it never was bliss of course, because the very mission of science was to lead us out of the wilderness of ancient fears, taboos and superstitions. But the promised land will never be won by science alone. The control of applied science is probably the greatest single problem for humanity today, and one might even go so far as to wonder whether the most penetrating social critiques, such as the theory of the class struggle and historical materialism, are not simply aspects of this basic question.

No doubt humanity has been facing it ever since the discovery of fire, but today it threatens its very existence. Everyone knows about nuclear power and the devastating possibilities of nuclear weapons,[55] but such apparently simple problems as the disposal of the radio-active waste from nuclear power stations are nightmares to those who worry about the social responsibilities of science.[56] Nowadays mathematical engineering is almost as dangerous, and the possibilities of 'artificial intelligence', and the vast computing machines which can and will be built, with their fabulous information storage and retrieval, are quite breathtaking. The privacy of the individual is now endangered, as are the rights of children to be taught by living teachers, and the safety of whole populations exposed to the danger of some electrical or mechanical fault when computers are harnessed to 'defence'.

The possibilities of biology and medicine are at least as challenging. My own professional background has made it natural for me to follow such developments. One of the largest fields in which they arise is that of generation, for this is the first time in human history that humanity is on the point of acquiring absolute control both of reproduction and infertility. All too soon we shall be in possession of means for controlling the sex of the human embryo. After this the sterilisation of whole groups might become a live issue. Ethical controversies have raged for years round contraception and abortion,[57] but problems are also raised by the new foetal medicine, which can detect grave abnormalities long before birth,[58] and by artificial insemination which is only by convention attached to infertile marriages.[59] Legal considerations and changes are lagging far behind the actual possibilities, such as spermatozoa banks, maintained from donors outstanding for physical or intellectual brilliance, and possibly several generations older than the receiving womb.[60]

Again, now that we know the chemical structure and coding of the DNA molecules which carry the instructions for making each new human organism, infinite possibilities are open for interfering with this hereditary material.[61] That would be biological engineering applied at the molecular level; one could envisage the insertion of an entirely new piece of chromosome, or the removal of another. Or one could produce hitherto unheard-of hybrids by substituting a mixed-cell nucleus for the original one of the fertilised egg itself. These may seem distant prospects, requiring enormous expenditure of money; but there has already been unexpected success in transferring genes (the hereditary units) from one lot of bacteria to another.[62] Certain viruses can pick up genes and put them into the bacterial nuclear systems. What if an antibiotic-resistant strain of bacteria were produced, which quickly spread all over the world like wild-fire, and decimated human populations? That this is a real danger has been shown by a self-denying ordinance achieved very recently in California,[63] where the scientists working in these fields agreed to establish a moratorium on such experiments, at least until more laboratories with adequate safety equipment and security become available.[64] Here there is a very tempting possibility, namely the possible insertion into plants of genes favouring the symbiosis of nitrogen-fixing bacteria, as happens in the leguminous plants today. If that could be arranged for the staple crop cereals it would be a gift to humanity almost as great as the gift of fire. What effect would this immeasurable increase in food production have upon the human race?

Medicine is also confronting humanity with almost insoluble problems.[65] The conquest of transplantation intolerance has already led to a great proliferation of organ transplants, and no doubt the surgeons in time to come will have access to whole banks of spare parts for human beings.[66] But transplantation studies go much further, for it is now possible to make chimaeras between animal species, since certain killed viruses makes the tissues stick together, and this could be used to unite human with animal tissues.[67] What is going to come of that? Ethical problems also arise in all cases where the treatment may be very expensive, needing elaborate machinery — for example, the kidney machines which dialyse the blood, and can keep a person going even though his or her kidneys are only able to function very ineffectively. Who is going to choose who gets the advantage of restricted techniques in short supply?

Again, much work is being done on the fertilisation and cultivation of human eggs *in vitro* up to the blastocyst stage before their reimplantation into a uterus to go on developing until term.[68] Aldous Huxley in his famous novel *Brave New World* visualised the isolation of totipotent blastomeres so as to reproduce many identical copies of low-grade human beings, and this is not at all impossible.[69] But there are other ways of effecting such 'cloning'. For example, nuclei from adult cells can take the place of the egg's own nucleus itself, so that a whole regiment of individuals with identical genetic material could be created. The question would then arise: do all human beings have an inalienable right to individuality? Such is the fix that Faust has got himself into, and the young suspect that they know why.

CHINA'S IMMANENT ETHICS

I could go on a lot longer, but it must be evident that humanity has never hitherto had to face anything like the tremendous ethical problems posed by the physico-chemical and biological sciences.[70] Now it is not at all obvious that the traditional ethics of the Western world, even with all its tomes of moral theology and casuistry, is the best equipped to deal with these problems, and certainly not on its own. Even within the sciences it is not obvious that the traditional modes of thinking of Western philosophy are the most adequate for the extraordinary and incredible events which go on in the world of sub-atomic particles, and indeed there are those, such as Odagiri Mizuho,[71] who are showing that Buddhist philosophy may give a good deal of help to the nuclear physicist which could not come from Western ideas alone. Here we can do no more than raise this point. What needs saying is that most of those who have worried quite properly about the control of applied science in the West have so far failed to realise that there is a great culture in the East which for two thousand years has upheld a powerful ethical system never supported by supernatural sanctions. This is where Chinese culture may have, I think, an invaluable gift to make to the world. Nearly all the great philosophers of China have agreed in seeing human nature as fundamentally good, and considering justice and righteousness as arising directly out of it by the action of what we in the West might call the 'inner light'. The Johannine light, perhaps, 'which lighteth every man that cometh into the world'.[72] Let men and women have proper training in youth, the right ideals, and a classless society which will bring out the best elements potentially within them.[73]

For the Chinese, then, ethics was accepted as internally generated, intrinsic and immanent, not imposed by any divine fiat, like the tables of the law delivered to Moses on the mountain. I should go so far as to say that never have the Chinese been more faithful to this doctrine, interpreting it in terms of selfless service to others, to people, than they are at the present time. *Wei jen min fu wu!*[74] This is my second fundamental point. If the world is searching for an ethic firmly based on the nature of man, a humanist ethic which could justify resistance to every dehumanising invention of social control, an ethic in the light of which humankind could judge dispassionately what the best course to take will be in the face of the multitude of alarming options raised by the ever-growing powers the natural sciences give us, then let it listen to the sages of Confucianism and Mohism, the philosophers of Taoism and Legalism.[75] Obviously we must not expect from them exact advice on choices arising from techniques which they would never have been able to imagine. Obviously, also, we are in no way bound down to the formulations they gave to their ideas in ancient feudal, or mediaeval, feudal – bureaucratic society – time marches on. But what matters is their spirit, their undying faith in the basic goodness of human nature, free from all transcendental elements and capable of leading to a more and more perfect organisation of human society. We need a renewed sense of justice and righteousness (*liang hsin*) and a profound understanding of what constitutes the fullest, healthiest life for man on earth. We need a firm realisation of the dangers to which science exposes us, and which we must at all costs avoid. The saying in the *Kuan Yin Tzu* book may well come true:[76] 'Only those who have the Tao can perform these actions – and better still, not perform them, though able to perform them!' Does not this echo strangely the narrative of the Temptations in the Gospel?[77]

CHINA AND THE MATTER – SPIRIT DICHOTOMY

Once again let us set off in a different direction. What else could Western society profitably learn from the Chinese tradition? What would have to come about in world society as a whole to allow of the integration, or 'oecumenogenesis', of Eastern and Western cultures? To begin with, there is one thing which seems to me to have been insufficiently discussed so far, and that is whether the extreme division of the world into spirit and matter, so characteristic of European thought, has been a good thing or not. It is certainly extremely un-Chinese, for we find in all branches of the history of science in China that there was a great reluctance to make this sharp dichotomy. China was profoundly non-Cartesian. I always remember a friend of mine at the BAs' table in Caius many years ago growing sarcastic at the expense of aberrant scientists like Oliver Lodge, and saying how 'these silly fools don't realise that you can't turn matter into spirit just by making it thin!'[78] Yet that is precisely what the Chinese were doing all through the ages. It is a curious thought that while matter in the West began as being very material, and got to be more and more ethereal as time went on, and the Bohr model of the atom came to be accepted; in China, on the other hand, *chhi*, which started as a very ethereal sort of *pneuma*, came to embrace even the most solid of matter by the time of the Neo-Confucian scholastics.

At all events, the disinclination of the Chinese to draw any sharp line of distinction between spirit and matter was at one with their deeply organic philosophy, and their psychosomatic medicine too, developed so long before Europeans came to the same outlook. Of course, in modern times there have been in the West many types of philosophy, such as that of A. N. Whitehead, which have done away with the sharp distinction, and a non-obscurantist organicism characterises the best experimental biology going on today. In the old *palaestra*, or gymnasium, where vitalism battled long ago with mechanism, a struggle continues between reductionists and anti-reductionists.[79] But it is rather shadow-boxing, because it must be perfectly obvious that, while the different integrative levels in living organisms only acquire their full meaning when related with the levels above and the levels below, the properties and behaviour of all living things up to the highest levels must be implicit *in potentia* in the nature of the protons and electrons that build them up. But of course if you only knew the protons and electrons, and had no knowledge of the highest levels of integration and organisation, it would be impossible to predict what those would be.[80] This is what I mean by non-obscurantist organicism. It fits into the traditional Chinese world-picture much better than it does to that of Europe, so bedevilled as always by traditional theological philosophy.

The sharp division between spirit and matter probably has as one of its most important correlates the equally sharp distinction between the relative value and significance of mental and manual labour.[81] It is evident that Chinese society today is making a great effort to overcome this, and to convince people that the use of the body is as important and health-giving as the use of the brain.[82] Mention has already been made of the recruiting of manual workers into the ranks of scientific observation, and by the same token it is considered highly desirable today that managerial, technological and intellectual workers should do a period now and then at the bench or in the field. No one who has any acquaintance with Chinese industrial production at the present time, even if only through printed or written accounts, can doubt that a tremendous revolution has taken place in industrial relationships.[83] Artisans, craftsmen and workers are encouraged to take initiative, bring forward inventions, tackle jobs co-operatively, and manage their own affairs within the framework of a great factory.

It is extremely interesting that the most enlightened sectors of Western industry have begun to follow similiar lines.[84] For example, it is common knowledge that two at least of the great Swedish firms in the motor car industry have completely abolished the assembly line, which in the days of Henry Ford certainly increased mass production but reduced the individual worker to a machine carrying out one simple task at a predetermined tempo along with other machines. Such jobs were dull and tiring, and destructive of the worker's self-esteem. Widespread disaffection showed itself, and continues to show itself, in the form of high labour turnover, chronic absenteeism, poor workmanship and even sabotage. Thus the principles of job simplification, repetition and close control are now giving way in many industries, even in the capitalist West, in favour of organisation in autonomous small groups shaped to fit particular jobs. Such groups can solve problems, learn from the problem-solving, and derive satisfaction from it. They can study, reason, evaluate and strive towards goals, while doing all the time not one particular repeated thing but a considerable variety of technical operations.[85] Now in so far as autonomous group production gives the worker more opportunity to influence his own job, to take on responsibility, to solve problems, and to advance his own development, as well as co-operating much more fully and joyfully with his mates, it is following methods and ideals which are typically Chinese.[86] This does not mean that the conveyor-belt is abolished, on the contrary it is still useful, but as a tool or means of transportation, not simply as a source of unvarying stress and strain. The delegation of power and influence to the manual worker is going to become quite normal and undramatic in all forms of developed society; and it springs from the basic principle that manual and mental work should not be regarded as diametrical opposites, but ideally combined in the same person. Anyone who has ever been a working scientist understands this. As the great Ivan Petrovitch Pavlov once said: 'I have done a great deal of mental and manual work in the course of my scientific life, and I think it is the manual work which has often given me the most pleasure.[87]

Here then is the third point I am making. Was it not a wise course the Chinese took when they avoided too sharp a distinction between matter and spirit? And perhaps it led to sociological wisdom too, the activity of the whole person in brain work and hand work; mind and matter no longer at loggerheads.[88]

CHINA AND THE CO-OPERATIVE MENTALITY

Just now the word 'co-operation' was mentioned, and again we have an extremely Chinese trait which is surely going to have a great effect upon the rest of the world. Mo Ti spoke about the 'ant attack', Ritchie Calder emphasised what could be done by 'a million men with tea-spoons'. In one of our volumes we illustrated a wonderful picture of a vast mass of men co-operating in hydraulic engineering work, the opening of a great new canal;[89] and I myself during the war and since have often marvelled at the way the Chinese could manage the large numbers of people needed for labour on airfield building, communications and hydraulic works. Here again the organic philosophy of China has much to teach the West. The atomic fragmentation of the competitive, acquisitive society destroys the personality. Competitive individualism only leads to alienation – quite unlike the cohesive atmosphere of the Chinese extended family, the guild or the secret society, or those who now join together in co-operative communes and industrial units.[90] There is of course a fundamental egalitarianism here, and my fourth thought is that it will be all to the good of the world if this sense of brotherhood spreads throughout mankind.

In what I am saying here, I am mostly commending Chinese ways of life and thought to the rest of the world, but I would be the first to admit that they do not yet solve all possible problems. Some acute Japanese friends, particularly Yamada Keiji, have pointed to a fundamental 'amateurism' which ran throughout traditional Chinese society, especially of the *literati*, and still today the Chinese ideal is to produce the 'all-round' person, the scientific, industrial, humanistic and aesthetic person, even though personally I might in some ways miss the religious element. But the problem then arises whether the Chinese will be able to conquer the 'commanding heights' of nuclear technology and similar achievements, if professionalism is too much subordinated to the all-round ideal. It is too early as yet to say. But of one thing I feel certain, namely that China will not produce those types of utterly inhuman scientists and engineers who know little, and care less, about the needs and desires of the average man and women. The 'new men and women' in China will solve this problem, aided by the infinite resources for development that are in the people themselves.

LOGIC IN CHINA AND THE WEST

Another thing I should like to say something about is the question of logic in China and the West. One of the earliest things I noticed about my Chinese friends, some forty years ago, was that so often they would not answer yes or no to my formulations, but something like 'well, not exactly'. Undoubtedly this was an outward and visible sign of a certain subtlety of thought which runs right through the whole of Chinese culture. From the point of view of the *Science and Civilisation in China* project it is obviously of the highest importance to elucidate what part logic and logical thinking played in relation to the development of the sciences in China.

At the present time two of our collaborators, Janusz Chmielewski from Warsaw and Derk Bodde from Philadelphia, are addressing themselves to this problem.[91] The results are as usual, somewhat surprising. First, it can be shown that formal logic was more fully and perfectly incorporated in the linguistic structure of Chinese than of any Indo-European language. Second, all the main methods of reasoning and forms of syllogisms can be found in Chinese philosophical and medical writings from the −4th century onwards. But, third, it is clear that no Aristotle, no Panini, arose in China to codify successfully the features of formal logic − Kungsun Lung and the Mohists attempted to achieve this, but because of the lack of interest of subsequent generations their writings were only imperfectly preserved and now have to be rescued from textual corruption.[92] It may have been precisely because of the profoundly logical structure of the language that the need for codification was never felt. Fourth, by the same token, the minds of Chinese thinkers were not mesmerised by abstract logic, so that full weight could be given to all kinds of nuances rising above the 'single-vision' of the 'either-or' dichotomy. Fifth, and lastly, comes the question, what relation did all this have to the development of science in China? The answer seems to be that it had no effect at all, whether in mathematics, astronomy, geology, physiology or medicine. Only the breakthrough to modern science did not take place, and it seems in the highest degree unlikely that that could have been due to the presence of formal logic in the West. For, as we all know, the founding fathers of the scientific revolution agreed with Francis Bacon's dictum *logica est inutilis ad inventionem scientiarum.*[93]

Speaking as one who was a working scientists for many years himself, I remember always feeling how unsatisfactory the '*A* or not-*A*' disjunction was. Of course it was obviously useful, indeed quite essential, for classification, but always as a preliminary sorting to be followed by further sortings. It was thus the basic tool of the taxonomist, no doubt. But for the chemist, the physicist or the physiologist, it seemed radically unsatisfactory because in Nature *A* is always changing into not-*A* as one looks at it, and the difficulty is to catch it on the hop.[94] In my time at Cambridge, no science undergraduate or research student ever dreamt of taking courses in formal logic, and over many years of attendance at tea-club meetings and lectures by colleagues, I hardly remember any occasions when people had to be criticised on account of logical fallacies. The premises and the statistical treatments were always much more important.

As for the history of Chinese scientific thought, the avoidance of rigid 'A or not-A' conceptions can be seen very well in the relations of the Yang and Yin. These two great forces in the universe were always thought of in terms of a prototypic wave-theory, the Yang reaching its maximum when the Yin was at its minimum, but neither force was ever absolutely dominant for more than a moment, for immediately its power began to fail and it was slowly but surely replaced by its partner, and so the whole thing happened over and over again. This is what Nathan Sivin has called the fundamental principle of Chinese natural philosophy, or the 'First Law of Chinese Proto-Physics'.[95] And even during those very short moments of time when Yang or Yin reach the height of their powers, still they are 'not exactly' all Yang or all Yin, because by an extraordinary feat of insight the Yang harbours a nucleus of Yin within itself and vice versa; and within that nucleus again there is an element of Yang, and so ad infinitum.

What China has to teach us here is, I think, that we ought to be much less rigid in our thinking and more flexible in our argumentation. This would mean that we would be more open-minded in many things, both scientific and social. We should be more ready to entertain ideas about possibilities hitherto unheard of, alternative technologies, experimental social groupings. In personal life we should be less conventional regarding human relationships, and more tolerant of all ways of life that do not break the law of love.[96] And of course in science we should never fear the new and utterly revolutionary. We should 'test all things', as the apostle says, 'and hold fast to that which is good'.[97]

CHINESE AND WESTERN ATTITUDES TO NATURE

The sixth and last major point which I would like to take up in this discussion is the question of the different attitudes to Nature held traditionally by China and Christendom. This is a subject which readily lends itself to vague generalisations, but nevertheless I think that something relatively concrete can be said.[98] Many half-truths (or even less) concerning it have been in circulation for a long time. For example,

F. S. C. Northrop wanted to characterise the Chinese approach as basically aesthetic in contrast with the scientific approach of Europe.[99] Or again, Fêng Yu-Lan once said that the Chinese philosophers had never sought to dominate Nature; it was themselves they sought to dominate.[100] If this sort of thing had been the whole story, it would obviously have been impossible to fill many volumes with the recital of the achievements of the Chinese in all scientific realms, from mathematics to medicine, over a period of some 2500 years. But there were great differences in the attitude to Nature, and here once again there is much for the world of today to learn from the Chinese tradition.

In the first place, it is evident to anyone who knows anything about Chinese civilisation that it did not have any well-developed theology of a creator deity. Unlike the thinkers of other early civilised peoples, the ancient Chinese philosphers did not give much credit to creation-myths accounting for the origins of the world. There were myths of organising and arranging gods, or demi-gods, but they were not taken too seriously. Chinese thinkers did not, in the main, believe in a single god directing the cosmos, but thought rather in terms of an impersonal force (*Thien*), meaning 'heaven' or 'the heavens' indeed, but here better translated as 'the cosmic order'. Similarly, the Tao (or *Thien tao*) was the 'order of Nature'. Thus, in the old Chinese world view, man was not regarded as the lord of a universe prepared for his use and enjoyment by God the Creator. From early times there was the conception of a *scala naturae* in which man was thought of as the highest of the forms of life, but nevertheless that did not give to him any authorisation to do exactly what he liked with the rest of 'creation'. The universe did not exist specifically to satisfy humans. Their role in the universe was 'to assist in the transforming and nourishing process of heaven and earth', and this was why it was so often said that humanity formed a triad with heaven and earth (*Thien, ti, jen*). It was not for man to question the way of Heaven nor to compete with it, but rather to fall in with it while satisfying his basic necessities. It was as if there were three levels each with its own organisation, as in the famous statement: 'Heaven has its seasons, man his government, and the earth its natural wealth.'

Hence the key word is always 'harmony'; the ancient Chinese sought for order and harmony throughout natural phenomena, and took this to be the ideal in all human relationships. Early Chinese thinkers were extremely impressed by the recurrences and cyclical movements which they observed in Nature — the four seasons, the phases of the moon, the paths of the planets, the return of comets, the cycle of birth, maturity, decay and death in all things living. *Fan chê tao chih tung*, as the *Tao Tê Ching* says, 'returning is the characteristic motion of the Tao'.[101] *Thien*, or Heaven, was more and more seen as an impersonal force generating the patterns of the world of Nature; phenomena were thought of as parts of a hierarchy of wholes forming a cosmic pattern in which everything acted on everything else, not by mechanical impulsions but by co-operation in accord with the spontaneous motivations of its own inner nature. Thus, for the Chinese, the natural world was not something hostile or evil, which had to be perpetually subdued by will-power and brute force, but something much more like the greatest of all living organisms, the governing principles of which had to be understood so that life could be lived in harmony with it. Call it an organic naturalism if you like; however one describes it this has been the basic attitude of Chines culture through the ages. Humanity is central, but it is not the centre for which the universe was created.[102] Nevertheless it has a definite function within it, a role to fulfil, that is the assistance of Nature, action in conjunction with, not in disregard of, the spontaneous and interrelated processes of the natural world.

Of course the ancient Chinese hunted and fished, but by and large their civilisation was always agricultural rather than pastoral; hence perhaps a more patient, less dominating, more feminine attitude to natural resources. The 'Man of Sung', who pulled up his sprouts in order to make the crop grow faster, was a standard laughing-stock for the peasant farmers of China through two millennia, accustomed as they were to a much more patient attitude towards Nature.[103] It is true that widespread deforestation occurred as the ages went by, but this should be put down to the pressures of social conditions, and warnings against it can be found in many texts, as in the *Mêng Tzu* book (Mencius),[104] or in the *Huai Nan Tzu*, protesting against the inordinate use of wood for firing metallurgical furnaces.[105] Warnings against depletion of natural resources are quite common in Chinese literature; another good example would be the action of that governor of Kuangtung in the later Han period, Mêng Chhang, who made the pearlers give a rest for a few years to the pearl-oysters, and afterwards guard against over-fishing.[106] Whenever anything could be done in accordance with Nature (this was the great *wu wei* doctrine of the Taoists) it was best to do it that way. For example, if water was wanted at 50 feet above the level of a river, it was much better to take it off by a derivate lateral canal some miles up-stream and follow the contours, rather than laboriously lift it by water-raising machinery at the spot. All this was not a 'passive' attitude to Nature, as some superficial minds have supposed; it was a profoundly right instinct that to use Nature it was necessary to go along with her. The Taoists would have applauded Francis Bacon's saying *Natura enim non imperatur nisi parendo* (Nature can only be commanded by obeying her).[107] Thus, to sum it up, there was throughout Chinese history a recognition that humanity is part of an organism far greater than itself, and by corollary a great sensitivity to the possible depletion, and pollution, of natural resources.

How different was all this from the feudal or imperialist domination of Nature arising from the Hebrew tradition. The People of the Book were never guided towards any restraint in the utilisation of those natural resources which God had provided for their use.[108] As long as modern science still lay in the womb of time, this lordship may have done no great harm, but once science had become airborne, as it were, then whole forests could be cut down every day to provide paper for the banal (and often vicious) printed matter of the popular press, while noxious chemicals, like organic mercury compounds or radioactive poisons, could freely spread about everywhere. Lynn White in his remarkable book *Machina ex Deo*[109] has revealed the responsibility of Christendom in inducing men to have an utterly possessive and destructive attitude to the rest of Nature, [110] a record only mitigated by the relatively minor voice of St Francis.[111] Western man will have to retrace his steps.[112]

Meanwhile today there is abundant evidence that the Chinese are conscious of the possibilities of pollution, which after all they have seen in horrifying forms very close at hand in Japan; and they are building into their industrial plants all kinds of arrangements for avoiding toxic or opprobrious effluents. Clearly this is something much easier to do in a socialist economy than when a private firm, or limited-liability company, is under the constraints of competitive marketing and profit-making. Modern science, too, is validating some of the characteristic Chinese techniques. For example, the use of human excreta as fertiliser has been a characteristic feature of Chinese agriculture for two thousand years. It was always a good thing in that it prevented the losses of phosphorus, nitrogen and other soil nutrients which happened in the West; but it was also a bad thing because it contributed to the spread of disease. But now in the light of modern knowledge of composting techniques it is quite easy to avoid the latter drawback while retaining the former advantage.[113]

One last thought which emerges from this discussion of attitudes to

Nature is that the People of the Book, and the West in general, have always been far too given to masculine domination. It seems imperative and urgent that the Western world should learn from the Chinese the infinite value of feminine yieldingness.[114] This is the message of the 'Valley Spirit' (*ku shen*) of the *Tao Tê Ching*.[115] Of course for the Chinese the greatest perfection always consisted in the most perfect balance of the Yin and Yang, the female and male forces in the universe. These great opposites were always seen as relational, not contradictory; complementary, not antagonistic. This was far different from the Persian dualism with which the Yin – Yang doctrine has often been confused. Indeed, the Yin – Yang balance might be a good pattern for that equilibrium between the forms of experience which we need so much, that harmony between compassion and knowledge-power. Here again then, there is something vital which the rest of the world has to learn from the Chinese tradition, if it is not to be torn to pieces by the interplay of intrinsic warring psychological factors, and external aggression against Nature and between men. The *ewig Weibliche* comes to us in Chinese dress, a Margaret – Gretchen, a Hsüan Nü, who can be the salvation of the world as she was of Faust himself.

It is high time that I came at last to my general conclusion. What I have been trying to urge is that in very many respects Chinese culture, Chinese traditions, Chinese *ethos* and Chinese human beings, those living today as well as those of all the ages, have contributions of outstanding importance to make for the future guidance of the human world. Nothing that I have been saying denies the 'Everlasting Gospel' of the two great commandments; but it is time that Christians realised that some of their highest values may be coming back to them from cultures and peoples far outside historical Christendom. The question is: what is humanity going to do with the Pandora's box of science and technology? Once again I should like to say: *Ex Oriente Lux*.

6
Science, Technology and Black Liberation

Sam Anderson

Forge simple words that even the children can understand. . .
words which will enter every house like the wind/fall,
like red hot embers on our people's souls.

Brother Jorge Rebelo (Frelimo)

What are science and technology? What are their origins? How do they function within the United States? What are the goals of Black liberation struggle and the political, economic and technical developments within the United States' science and technology?

These are some questions that exist at this point in our revolutionary development. There has been very little effort in the United States to answer these questions, fundamentally because science and technology have been traditionally viewed by the white left as an appendage to the socio-economic system of capitalism with, until recently, an apparent apolitical function. However, today science and technology are put into their proper perspective, a necessary political agent for the development of monopoly capitalism and imperialism, and maintenance and rationalisation of white racism (personal and institutional).

> Science today is property, and therefore, like all property, it is used for the benefit of those who own it. In the USA and in other imperialist nations, the major part of scientific effort is dedicated to the twin purposes of (1) extraction of profits and (2) the maintenance of the control which permits that extraction.[1]

> The key innovation is not to be found in chemistry, electronics, automatic machinery, aeronautics, atomic physics, or any of the products of these Science-technologies, but rather in the transformation of Science itself into capital.[2]

Specifically, within Afro-America — in spite of one's political position — science, technology, scientists and technicians are consciously or subconsciously placed on an envious pedestal: '(s)he is heavy because (s)he is into the sciences'; 'Man, the sciences, engineering and medicine are deep or heavy subjects.' That is, more complex and disciplined than the subject matter or work in the bourgeois humanities and social sciences. In short, because of our fear, mystification and envy of the natural sciences, technology and scientists, our ideological development and day-to-day struggles are similar to and have virtually the same results today as Black reconstruction did in 1874.

But first let us set a political and historical framework from which we can begin to understand what is and what needs to be done.

WHAT ARE SCIENCE AND TECHNOLOGY?

J. D. Bernal — a British Marxist physicist and science historian — has written a most comprehensive and readable analysis of the evolution of science and technology in human history called *Science in History*. Rather than give a pat 'Webster's Dictionary'-type definition of science (and technology) Bernal states:

> science has so changed its nature over the whole range of human history that no definition could be made of it. Although I have aimed at including everything called science, the centre of interest. . . lies in natural science and technology because. . . the sciences and society were at first embodied in tradition and ritual and only took shape under the influence and the model of the natural sciences. . . . Science stands as a middle term between the established and transmitted practice of men who work for their living, and the pattern of ideas and traditions which assure continuity of society and the rights and privileges of the classes that make it up.[3]

Science, then, becomes both an ordered technique and rationalised mythology.[4] Given this reality and given the reality of its complete fusion within a contemporary capitalist or socialist society, we need to perceive science (and technology) as (a) an institution, (b) a method, (c) a cumulative tradition of knowledge, (d) a major factor in the maintenance and development of production, and (e) as one of the most powerful influences shaping beliefs and attitudes toward the universe and humanity.[5] It is within this context that we will use 'science and technology' or just the term 'science'.

THE ORIGINS OF SCIENCE

An essential factor in aiding us to overcome the psychological barriers against understanding and working with the sciences is to place science in its correct historical context. Black folk have to struggle against a double psychological barrier: science as divine and mysterious, and science as non-black in the socio-historical sense. Keep it in mind that there has been no societal development without scientific and technological development. One of the earliest revolutionary scientific developments that moved mankind in general and our African ancestral brothers and sisters in particular, into a new social order (the city-state) was the development of iron smelting in Zimbabwe (Rhodesia) at least 40,000 years ago.

This scientific and technological achievement could not have occurred without other scientific achievements in the area of agriculture. It implies that farming (a technological achievement, primarily developed by women) was necessary for the development of a more stable and non nomadic society. This recent discovery of a 40,000 year-plus old Zimbabwe iron smelting,[6] implies that scientific and technical information moved from Southern Africa to Northern Africa and then on to Europe. This would completely shatter influential racist historians like Arnold Toynbee and his *A Study of History* where Egypt is depicted as a white nation and the rest of Africa a mere savage pre-historic footnote.

More important than the necessary refutations of racists is a proper analysis of the developments and dynamics of science within those societies before the onslaught of European slave trade, colonialism and neo-colonialism.

The African continent reveals very fully the workings of the law of uneven development of societies. There are marked contrasts between the Ethiopian empire and the hunting groups of pygmies in the congo forests or between the empires of the Western Sudan and the Khoisan hunter gatherers of the Kalahai Desert. Indeed, there were striking · contrasts within any given geographical area. The Ethiopian empire embraced literate feudal Amharic noblemen as well as simple Kaffa Cultivators and Galla pastoralists. The empires of the Western Sudan had sophisticated, educated Mandinga townsmen, small communities of Bozo fishermen, and Fulani herdsmen. Even among clans and lineages that appear roughly similar, there were considerable differences.[8]

Then we have arrested development and underdevelopment in African (and Asia and Latin America) for capitalist development in Europe and the United States. The social factors which congeal to determine just when a society makes that dialectical leap from small-scale craft technology to equipment designed to utilise nature in such a way that labour becomes more efficient are varied. One primary social factor is the demand for more products than can be produced by hand. Hence, technology and technicians respond to a clearly defined social need, clothes for example. When European cloth dominated the African market, African producers were severed from the increasing demand. What followed was that craft producers either dropped their skills because of the onslaught of cheap, readily available European cloth, or they struggled along subsisting on the same slow manual weavers to create styles and pieces for very localised and subsequently small markets. Therefore we witness *continental* technological arrest, stagnation *and* regression. The use of and the products of European capitalist technology *forced* thousands of Africans (and Asians and Latin Americans) to forget not only the complex techniques of their ancestors but even the simpler ones. When one understands the centrality of smelting to development, one can see that the crucial technological regression for Africa occurred when traditional iron smelting was abandoned.[9]

In short, science and technology during the pre-capitalist periods were advancing on a global scale. But European and US capitalism made scientific and technological development occur only in Europe and the United States.

Hence previously independent societies throughout the Third World today become dependent societies trapped at best in the eighteenth and nineteenth centuries. The more scientific achievements occurred within Europe and the United States and the more efficient and exploitative capitalism became, the more the burden of capitalism was placed on Black and Third World people.

Looking within the growth of capitalism, we find that during the primitive capitalist stage science was virtually non-dependent upon the capitalist for financial support. On the other hand, the mere existence of a primitive capitalist society allowed a few individuals from merchant families or other non-working-class groups to pursue the time-consuming work of scientific inquiry (as opposed to a hit-and-miss or non-experimental inquiry).

We need only to look at the socio-economic roots of Isaac Newton's *Principia* to understand that this work evolved out of the needs of early capitalist development and its attendant technology. In short, *Principia* was not, and could not be, a work of 'pure' science by a value-free or 'neutral' scientist.[10] As the capitalists forced their way across Europe, Africa, Asia and the Americas, their system demanded a larger output of goods, which in turn demanded larger, more efficient machines to extract and transport the raw materials. Thus, for example, the invention of the cotton gin initially meant a growth *both* in African slave labour in the South *and* European immigrant labour in the North, and in Britain the forcing of peasants into an industrial proletariat.

But, as the cotton gin paved the way for more mechanised agriculture, it also aided in the forcing of black Americans out of agriculture at a time when European immigrant workers were occupying the production jobs in northern industries. Again we see an apparent contradiction: as capitalism provided/demanded more leisure time for the bourgeois Europeans, it encouraged the development of scientific inquiry, and it used the results of this methodology for its own ruthless and grotesque growth:

> In imperialist countries, the scientific venture is devoted, for the most part, to the development of military technology, to mass extermination, and to fascistic control of the behavior of society as well as of the individual. The objective benefit that humankind might gain from scientific work is of secondary consideration.[11]

Science, then, has become an inseparable part of capitalist development, and capitalist development has become an inseparable part of science. But just as capitalism as a system/civilisation is digging it own grave, we have a withering away of capitalism through its prodigal son – modern science:

> The old concept, which goes at least as far back as ancient Egypt, of an educated elite, and a mass of illiterate peasants and workers, is bound to disappear, indeed is disappearing already. The archaic methods of control of production and consumption sanctified in the code of free-enterprise capitalism, which has become monopoly capitalism, will have to make way for planned production and to make more and more use of mathematical and computational methods. In words, science implies socialism.[12]

But let us be very clear: that science implies socialism does not mean that there will not be, or should not be, a struggle between the tiny ruling class and the working masses of humanity. It does not mean that

all Black folk in the Americas and Africa have to do is try and become
scientists and then somehow bring about socialism because they are
scientists. A people's struggle and the struggle for a people's science are
intimately bound up, as victoriously demonstrated by Cuba, North
Korea, North Vietnam, Mozambique, Guinea-Bissau and the People's
Republic of China.

BLACK AND THIRD WORLD DEPENDENCY AND TECHNOLOGY

When we look at science and technology as they function within Third
World societies we see two contradictory processes. The first is a
liberation process. Science and technology have freed large sections of
societies from the killing extremes of labour and diseases. The other is
an oppressive and exploitative device. The American and European
capitalists have invented mechanisms so subtle and sophisticated that
they come disguised as 'progress' and 'development'.[13]

> For this reason, science is like a smoke-screen: while its force appears
> to be directed at the resolution of the most urgent problems of our
> peoples, it makes these problems more numerous. It covers up the
> social roots of 'technical' problems. In the rhetoric of 'harmony' it
> enshrouds the reality of imperialism.[14]

The good that contemporary science and technology can do is over-
whelmed by the imperialists' quest for raw materials, new markets and
peace at home through mass material consumption. When we turn to
black America on this matter there is no improvement. In fact, black
America is so technologically backward relative to its position inside of
the world's most technologically advanced nation that we are more
backward than the newly independent Republic of Guinea-Bissau!
More precisely, black America is more technologically dependent upon
capitalist America than is Guinea-Bissau. Because of the type of socialist
society the liberation struggle is creating, in Guinea-Bissau, we will see
science and technology evolving *from* and for the masses. We will see
less reliance upon the Western form of industrialisation: products-as-
trinkets and pacifiers for bourgeois consumption.

But inside the United States, inside black America, the dominance of
capitalism's technology and racist culture deforms most attempts of
our scientists and technicians to create in a human-orientated way.
Blood plasma, like Chinese gunpowder, initially non-military, became
an essential piece of war equipment so that modern racist-capitalist

wars could be fought more efficiently and victoriously (for the capitalists and 'manifest destiny'). Technological industrialisation in black America has oppressed and exploited Black folk in at least three ways outside of the economic super-exploitation:

(a) Producing goods for the white multinationals and military to be used against ourselves and our Third World brothers and sisters;

(b) Producing goods for pacification and bourgeois consumption; and

(c) Producing goods that are outdated and/or irrelevant.

Moreover, these types of production only force us to become even more dependent upon the whims of dominant white America. Like the countries of Africa, Asia and Latin/Caribbean America, we must be able to judge how desirable a product is by the number of linkages to other production activities within the economy: 'Where this is not accomplished and productive technology exists as an enclave, local production is not effectively stimulated and the diffusion of whatever technical advantages may be contained is operationally sterilized.[15] This means to black America (and to Third World nations) growth without development — dependency and a more efficient imperialist machinery.

Let us look at this in more detail. If we are to ever see Third World people break away from the imperialist strangle-hold and advance to a complete liberation, we will all have to confront Western science and its promised technological miracles just as our ancestors should have confronted the christian missionaries with their talk of miracles and heavenly salvation. This confrontation will have to be face to face, with the same resolve and firmness which our Vietnamese brothers and sisters used when they confronted the naked brutalities of the United States' technological god.

We will have to remember that capitalist technology and its accompanying technical advisers and experts enjoy a lot of publicity and are always up for sale. It was all designed to serve an individualistic competitive consumer society in the metropolis and imperialist domination elsewhere. Throughout history, the way in which Third World people have come into contact with technology used by Europeans and North Americans has always been through wars and capitalist expansion determined by the desire for profits of a few concentrated interests and de-emphasising the human element in science and technology.

Science and technology is not just hanging out in space waiting for the natives to be civilised enough to get hold of it piece by piece following the white masters' advice while remaining at his service. Technology is nothing more than a part of the capitalist package of wares that the Western world uses to exploit and control the Third World. It was manufactured in the heads and institutions of the European bourgeoisie to serve its greed and its requirements for more efficiently exploiting its working class. Technology cannot be separated from the material base at which things are produced through labour; what Marx calls a 'mode of production' determines which style of science will be favoured, which technological innovation will be judged 'useful.' Under capitalism the 'valuable' innovations are those which bring profits to the private owners of the means of production, that is to the capitalists. When the capitalists looked abroad and built empires, they again decided which technical tools — if any — would be 'useful' to have their peanuts raised, their cotton picked or their bananas carried into the big white boats. Imperialists have no special mystical belief in technology, however, when their economic interest is at stake. They may even decide to kill existing indigenous technologies if that is good for profits; thus, just as the looms of Africa were forced into idleness, so were India's as Britain developed mechanically produced cloth in its ruthless, child-laboured sweatshops in Manchester. So, today, when the white man turns around and appears to be generously offering his investments, technical gadgets and know-how as assistance, as the cure-all for what he calls under-development, we had better assume that he has something good (profit) for himself in mind.

What we will accept of it all, what can be salvaged of the technical 'successes' of that decadent system, must be carefully chosen and re-shaped, just as the Cuban revolution salvaged and reshaped the military camps of Batista's buildings.

At this moment the Third World masses, in their struggle for socialist progress, are struggling for themselves, by themselves. We know that the white man's 'progress' which drops shiny tin cans on the moon will not give us food. Less obvious, but more important to realise, is that the so-called Green Revolution and its transnational agribusiness management will not feed us either.[16]

SCREENING ANY PEOPLE'S CONTROL

Inventions are not made in a vacuum or from the vacuum; 'new' ideas do not fall from the sky; they are part of a world of attitudes, ways of thinking and previous ideas about reality; they all merge together into what is called an 'ideology'. The calculus of probability was invented to help a European prince win at gambling! Today, in the citadel of world imperialism the choices of technologies to develop are still made by the princes of science, who sit on the President's Science Advisory Committee and outline the virtues of fragmentation bombs and lethal gas. The reigning ideology allows bourgeois scientists to feel at ease in such a role and to identify their work on weapons with the pursuit of science itself; thus the distinguished chemist Louis Fieser, on whose textbook many chemists were brought up, and who happens to be the inventor of napalm, wrote a book which describes the development of this particular weapon under the distant, depersonalised and universal title, *The Scientific Method*.[17]

In contrast, our world is the one in which, for example, the struggle of Frelimo (the Front for the liberation of Mozambique) has taken place, in which liberation has been achieved on the Mozambicans' own terms. As Samora Machel stated to the liberation fighters: 'Education must give us a *Mozambican* personality which, without subservience of any kind and steeped in our own realities, will be able, in *contact with the outside world*, to *assimilate critically* the ideas and experiences of other peoples, also passing on to them the fruits of our thought and practice.'

So there is an absolute need to discriminate, to choose, to exert what we shall call an over-all 'cultural screening'. By cultural screening we mean the right to choose a style of technological creation that emanates from and is useful to the people, a technology emphasising the satisfaction of public or mass needs as opposed to individual consumption (that is private cars, changing styles of clothing, gadgetery, and so on). Screening, however, cannot just be a choosing of what to let in. More important is how the usage of new techniques is kept under the control of the working (peasant and proletarian) population. Furthermore, that control, that taking of matters of choices among technologies and styles of development into the people's hands, is only meaningful if the people themselves are capable of exercising judgements and making decisions, if the adaptation, the introducing of technology is made part of the over-all political struggle, consciousness-raising and educational campaigns. What to do about technology must be as central and debatable as how to change the judicial system, the holding of trials and instituting a people's justice. A people's technology is only meaningful if the people at large are involved in discussing it and shaping what it will be. This is being successfully accomplished, as one can readily observe in China, North Vietnam and North Korea.

Third World countries need to be aware that even in the scientific and technological areas there needs to be a close scrutiny of the politico-cultural effects of those technologies which were originally developed in a capitalist framework. Development will not come with a single transfer of US Natural Science Departments and Schools of Engineering or Architectural Institutes. Nor can one transfer the practice and concepts of US medical 'treatment' on to an underdeveloped country without separating the wheat from the chaff: the humanitarian aspects from the anti-humanitarian capitalist aspects; the part which is directed at helping the sick and the part which uses the sick as research fodder.

For these reasons we maintain that a continuing control over technology's orientation on the part of the peasants and proletariat is absolutely necessary. Otherwise, if technology, under the excuse of being 'neutral', is accepted at face value, an elite of technocrats will play a vital exploitative role in determining the socio-political atmosphere. Such an elite will always favour a capitalist-oriented mode of development, which naturally is in its class interest. The struggle to politicise the work of scientists and technicians and the struggle to create a people-controlled science are part of the over-all class struggle, even during the phase of socialist construction.

STRUGGLING OUT OF DEPENDENCY AND UNDERDEVELOPMENT

According to the ideology of Western civilisation, the way of presenting the current situation in the Third World — as pointed out above — begins by ignoring history. Thus the world appears as a god-given set of nations with some that happen to be more developed than others. There are no causes for this situation. It is never presented as part of historical process, and therefore all solutions offered for development are based on mimickery of the 'advanced' countries. 'Western thought' takes for granted that capitalism will always exist and that Third World countries will always be behind although they can get a little closer through some fatherly development programme. The over-all vision is that since the Industrial Revolution was good for Britain, imitative technical development will be good for the Third World and will 'close the gap' between rich and poor. The necessity of technological manufacturing becomes an overwhelming belief, a development status symbol. But one ends up with production of luxury goods for the consumer in the developed world. This orientation is the logical continuation of the historical situation from which today's Third World results. The basic dialectical confrontation remains the same: to have colonised people, one needs to have oppressors. The officially fashionable question, 'how can developed nations help underdeveloped ones?' must be recognised as a false issue until it is put into historical perspective by first asking Brother Walter Rodney's question: How [did] Europe underdevelop Africa? In 'benevolent' development plans, the Third World, in fact, has only provided cheap labour and in many cases natural resources, but has been led to believe that it did develop industrially because it has some modern (or not so modern) plants to show. Even if some of those plants produce goods of use to the masses of the country itself (like shoes), the machines remain foreign-built and therefore dependency continues via spare parts and patent rights, and any 'modernisation' plan would also have to come from abroad. Taking seriously the need to build an on-going people-controlled technology leads us to face the fact that it is the people themselves who must do the development. Just a few foreign-trained specialists will not do. To keep matters in their own hands the people have to rely upon their own forces. This policy has been carried out by Cuba, confronted by the blockade imposed by the United States. It has been adopted by the peasants and workers of China under the name of self-reliance and by the Korean people under the name *Juche* or self-sustenance. Kim Il Sung declared in 1967:

> only when Juche is established firmly in scientific research work, is it possible to bring the initiative and talents of scientists into full play, to accelerate the advancement in science and technology and develop our economy faster by relying upon the resources of our country and our own techniques. Scientists and technicians should concentrate their efforts on the research work designed to promote industrial production with domestic raw materials, tap those raw materials which are in our country and produce substitutes for the raw materials we do not have, and expedite the technical revolution in conformity with our actual conditions so as to free the working people from arduous labour as soon as possible.

Self-reliance encourages the development of indigenous techniques and aims at furthering the creativity of indigenous technicians.

On a national scale, self-reliance implies establishing an inventory of a country's own resources. This necessary inventory was never carried out under colonial rule because the coloniser was usually interested in sapping the country of only those natural resources vital to the metropolitan country's industrialisation or providing quick cash returns. Today, the neo-coloniser — the native bourgeoisie — follows the same dictate: What's good for the metropolitan society is 'good' for the neo-colony. For instance, the Ivory Coast and Nigeria were formerly known for their palm oil and their cocoa; today their neo-colonial bourgeoisies are satisfied with cashing in on tourism and crude oil respectively. Meanwhile, Jamaica and Guyana remain the paradise of US — Canadian aluminium companies. In fact, all countries in the Third World, except for the socialist-orientated ones, follow this degenerate pattern.

Having such an inventory allows the peasants and proletariat to use what is nationally available to produce energy, food and building materials. Once the existence of these basic natural assets is publicised there is stimulation for co-ordinated and planned action by the people at the local level. There emerges a national reality within which initiatives for development can be taken. However, means must be found to maintain popular political control over these riches. What is creatively done must evolve through peasant and urban worker leadership within the political party; otherwise there will be no guarantee against local or state level capitalist sabotage or bourgeois sell-outs.

What are some of the advantages resulting from a mass-based self-reliant way of using technology? First, one should not forget that self-reliance may initially impose great demands upon everyone. Witness those participating in land-reclamation campaigns in Algeria, Cuba or Somalia, being transported — standing shoulder to shoulder — in trucks in the choking clouds of dust from unpaved highways. But the people involved were and are bound together by their common work experience preparing them to successfully confront future struggles demanding their technical innovativeness. Again Samora Machel:

> Releasing the masses' sense of creative initiative is an essential precondition for our victory and one of the chief purposes of our struggle. If the masses are to exercise the power so dearly won, they must display initiative. Colonial oppression, tradition, ignorance and superstition create a sense of passivity in man which stifles initiative.
>
> To create a sense of initiative is also to create a sense of responsibility and to make [the masses] feel directly concerned by everything related to the revolution, to our life.

Scientific and technological development and the preparation of scientists and technicians must be governed by the needs of the over-all political economy directed by the socialist ideology of self-reliance. Initially such technical development has to rely upon indigenous techniques. In turn these will be freed from the stagnation imposed by colonialism and from the contempt emanating from the neo-colonial bourgeoisie, drunken on mimicking the West.

Indigenous techniques which have evolved over many generations contain within themselves the answers to many specific local demands and problems. The looseness of the weave of some African cloth, for example, is directly related to clothing needs in tropical areas. The treatment of the fibre has evolved into a process which aids the resistance to mildew. Besides, such unpatented technical realisation in indigenous production is also an art involving the participation of various people in co-operation at the village level, from weavers to dyers. This is to a large extent a communal process of production which gives the people a sense of control.

In sum, technological oppression and dependence blocks us from developing (*a*) revolutionary black scientists, (*b*) scientific and technological alternatives, and (*c*) scientific education among the masses of Black folk. Given this reality, then, what is a 'revolutionary' black scientist, what are scientific and technological alternatives, and what is scientific education among the masses? Ultimately we have to struggle with the function of science and technology in the development of a twenty-first century revolution. Hence what follows is what Brother Amiri Baraka calls 'answers in progress'.

BLACK AMERICAN SCIENTISTS FOR THE PEOPLE

The majority of the few black scientists and technicians that we have are systematically and pathetically caught up in the 'special nigger' syndrome. Most of us see ourselves as scientists who happen to be black; the elite and thoroughly enlightened class. Individualism and professionalism are the norm. Not only the number of degrees are important but where and in what field are vital questions. In short, the typical black scientist or technician is staunchly middle American. We loyally believe in the myth of science as apolitical and 'objective', as something pure and above humans. We try to believe in this myth with such emotional force that we have become the most alienated of Blacks from black America. And yet with all of our conscious and unconscious rejection we are still the missing link, the ex-slave, the nigger.

This elite group, which constitutes much less than 10 per cent of the 624,400 black professionals and managerials, is for the most part irreversibly fixed in its plastic and crass bourgeois attitudes. The question then is, 'how do we struggle to develop a cadre of brothers and sisters who see the need to utilise modern science and technology as weapons and constructive tools?'

Above all else, if we are about to struggle to be serious scientists, then we must recognise the intellectual imperialism of the United States' ideology of an industrialised, bureaucratised and technocratic science: a social process which severs feeling from thought and which, in turn, splinters scientists into tunnel-vision specialists who cannot and must not see the whole. This is an alienation from people such as contemptuous objects for experiments. And the people respond with feelings of inferiority and dependency. . . or respond with revolution. But, for us:

> If we do know that there exists a science which is imperialist in its uses, its organization, its method and its ideology, there must exist, and in fact there does exist an anti-imperialist science. It is still in its infancy, and it takes different forms, according to the conditions it is found in. In colonial countries, dependent countries, or imperialist countries, it begins by exploitation; we denounce the use of science's name in the new pseudo-scientific racism; we denounce the conversion of science into a commodity and of our universities into corporate offices. From denunciation we move to active criticism: we look for means to put our scientific knowledge at the service of the people, and therefore as an instrument of revolutionary national liberation movements.[18]

Over the past four years many brothers and sisters within the National Black Science Students Organization (NBSSO) have been trying to resolve this question by organising, recruiting and politically educating black students. It is from the hard work of these brothers and sisters that we begin to see the fruition of ideas and actions outlined below.

Blacks in America have reached that critical point where they are the leading and most revolutionary force within the American proletariat. The development of history has placed us within, if not closer to, the major industrial metropolitan centres. But, at the same time, industry is becoming more and more dependent on scientific and technical innovations while we are becoming a less skilled, a less scientific people. Our method for developing a revolutionary ideology and our relationship as scientists to the modes of production have been anti-scientific. To resolve this problem, the primary intention for the formulation of a Union of Black Scientists and Technicians (UBST) should be to struggle with our fellow scientists and technicians (from semi-skilled workers to engineers) around developing a scientific methodology, and hence a cadre of revolutionary scientists and technicians:

> We challenge the system of training which tries to continue producing obedient experts. We are beginning to develop a new science on behalf of the whole of technology and society — an integrated science which refuses narrow specialization and idiot realism. We repudiate hierarchical-classist structures in order to search for forms of collective work and more democratic forms in research as well as in training. We repudiate the mystification of a science reinforced by a specialized vocabulary and we launch a campaign to popularize science. As scientists and revolutionaries we unite with anti-imperialist scientists of the world and with popular movements of our countries.[19]

Let us look at one proposed model of UBST. It proposes the following functions of the Union of Black Scientists and Technicians:

(1) First and foremost is the dire need to support the NBSSO's effort to recruit more Blacks into the sciences. Both organisations could help in politically educating our science students.

(2) The Union should present an anti-racist, anti-capitalist stance to both the sciences and world developments.

(3) UBST should formally work with the Scientists and Engineers for Social and Political Action (SESPA) — a predominantly white radical science collective — on major projects of crisis proportion, for example the fuel crisis.

(4) The Union should focus on about three or four major projects which have international implications. For example, the research and development of alternative energy sources for Africa and the Caribbean could be undertaken by brothers and sisters with the Union. We have only scratched the surface on how to cheaply and efficiently tap solar energy, so plentifully available in Third World countries.

(5) UBST must be in the forefront of science curriculum development from kindergarten through to the universities. One of the crucial problems facing black America is that of creating a tradition of dedicated, politically aware and numerous scientists and technicans. We have to break down psychological and skill-deficient barriers by

re-analysing the whole of science and technology and how it has been taught in the United States.

(6) The Union should give technical assistance to those community organisations that are fighting against the mind control and spying onslaught of cable television. This is a crucial issue that has stunted political and educational development in black America because we are technically unprepared to struggle on the level of the enemy.* The cable television situation is extremely serious. The government today has the technical know-how and the political necessity (creeping fascism to protect imperialism) to make the following scenario a reality:

the government, in a study of 1000 known community organizers has developed a profile of their viewing habits over an extended period of time. Computers then scan the output from tens of thousands of cable TV sets over a period of time. Individuals whose viewing profile are congruent or largely overlap with those of the activists are then selected out for 'further study.' In this way, 'troublemakers' could be identified even before they ever began to make trouble.[20]

(7) Alternative research into the genocidal functions of the birth-control system is another vital struggle for the Union. For it has been scientifically shown that the so-called population explosion is a myth of, and a fear by, the capitalist. We must relentlessly struggle with our brothers and sisters to understand the myth of the population explosion as clearly as Brother Eduardo Galeono:

The United States is more concerned than any other country with spreading and imposing family planning in the farthest outposts. Not only the government, but the Rockefeller and Ford foundations as well, have nightmares about millions of children advancing like locusts over the horizon from the Third World. . . . Its aim is to justify the very unequal income distribution between countries and social classes, to convince the poor that poverty is the result of the children they don't avoid having, and to dam the rebellious advance of the masses.[21]

(8) Organised black scientists and technicians will be compelled to expose to the masses of Black folk the neo-eugenic movement for

*The enemy is: The Federal Communication Commission, the American Association for the Advancement of Science, Dr Amitai Etzioni, Director of the Center for Policy Research, New York City Office of Teleeommunication Policy, and so on.

what it is: a racist pseudo-science. The Union must point out the class (bourgeois) basis of the neo-eugenic thrust as well as the gross misuse of science. Most of the work on exposing the neo-eugenicists has been done by a handful of white radical scientists and activists, but this fundamentally technical work has not been translated into a non-technical language and disseminated throughout the Black and Third World communities.

(9) The Union should struggle for a people's implementation of computer-related material. It is clear that black America is one of the most alienated sectors of the American society from computers; and yet computers function as necessary mechanisms for oppressing and exploiting us. On the other hand, it is obvious and necessary to see that computers can play an important role in our struggles against racism and imperialism.

The Union of Black Scientists and Technicians then is not a union of Black folk reacting to the fact that the white scientific groups do not want to integrate at the level of policy and power. Rather it is an organisation of people who assume and know that within the complex machinations of racism and imperialism there are supportive white racist science organisations dedicated to the perpetuation of global imperialism . . . dedicated to the perpetuation of black oppression, exploitation and genocide. The union is also an organisation that must subordinate itself to the necessary Black United Front and ultimately to the revolutionary vanguard multinational party that will guide and develop the proletarian liberation struggle within the United States. We must recognise that the UBST is fundamentally petty bourgeois in its composition and the task — monumental as it may be — is to convert as many of our sisters and brothers to the side of the working-class struggle to overthrow racism and capitalism and struggle to create a socialist United States. We cannot see ourselves as leaders, for that only leads us down the road of technocratic fascism.

But, meanwhile, what does an individual black scientific worker do? As one of America's peculiar beings (Black and a scientist or technican), we suffer under a barrage of jive propaganda from government, industry and university sources about the fame, money and privileges we can have because of our peculiarity. We have to recognise these 'niceties' and 'privileges' as the carrot of the racist-capitalist. The process by which we transform ourselves into Black, conscious and revolutionary beings is an extremely painful and demanding process. Thus the suggestions that follow are few and broad — but they are things that can be done by every one of us who wants to dedicate his or her life and mind to the liberation of black America through the liberation of America.

Read and discuss among your co-workers, family and friends (a) *How Europe Underdeveloped Africa* by Walter Rodney (b) *Science in History* by J. D. Bernal, (c) *Science for the People* magazine, (d) *Accumulation of Capital on the World Scale*, (e) *Labor and Monopoly Capital* by Harry Braverman, (f) *China: Science Walks on Two Legs* (Science for the People).

Set up small 'Science for Blacks Study Groups' that would discuss current issues and the development and politics of science aid projects that would necessitate the use of the group's skills.

Encourage more sisterly and brotherly relationships between the worker and the 'professional' in a technical area. After all, to the owners and other Whites in the industry, we are all first and foremost niggers.

Each of us should encourage at least two sisters or brothers to pursue the sciences for Black liberation. Help them find a college and stay in school.

Help construct the Union of Black Scientists and Technicians by supporting every stage of its creation through constructive criticism and work.

In 1968 Black Panther brothers and sisters shouted 'the spirit of the people is greater than the man's technology'. They were not being anti-science. Those brothers and sisters — in the midst of a black tidal wave — were saying that out of the spirit and will of our people to struggle for liberation there will come a science and technology for and from the people — in spite of the odds against us.

7

Ideology of / in Contemporary Physics

Jean-Marc Lévy-Leblond

APPRAISING IDEOLOGY IN SCIENCE

Today one could say of ideology, what used to be said of the devil: his best strategem is to make us disbelieve his existence. Thus, as a general rule, ideology has to obscure itself to achieve its prime effect. This is why its presence, even when recognised, is not easy to bring out. Scientific activity, in particular, often describes itself as typically non-ideological, invoking the myth of 'scientific objectivity'. That the denial of the role of ideology in science is simultaneously an admission of its importance, is not a new proposition. Still, it remains to be proved in any particular case, and the nature of such a proof very much depends on the meaning given to the word 'ideology'. I will not proceed here to define rigorously the concept as I use it. Let me stress rather that I am not so much concerned with ideology in general and in the abstract as with definite forms of it; I am interested mainly in the development of contemporary physics under modern capitalism; it is the role played during this development by bourgeois ideology (here and now the dominant one) which I seek to understand. Most of the analysis here will be devoted to this task. This is not to say that I claim this essay is written from a non-ideological point of view based on a supposedly super-(or supra-) scientific philosophy. Indeed, if I believe it possible to identify ideological effects (and major ones) in and of science, I think it meaningless to attempt to separate the scientific wheat from its ideological chaff. Quite the reverse, I argue that ideological elements cannot be separated from 'scientific' ones since they coexist *within* the

* This text has benefited from some of the remarks and criticisms offered by E. Balibar and P. Roqueplo.

real practice of science, which ideology underlies and conditions. None-theless I do not use 'ideology' in a systematically depreciative sense; rather, I wish to take into account the nature (bourgeois — or not) of the dominant ideology operating in a particular historical situation This is to say that another science requires another ideology, not the (impossible) elimination of ideology.[1] Only an 'ideology for (that is of and by) the people', will help to build a 'science for (that is of and by) the people'. This also means that such a task essentially lies ahead of us: we scientists, as radicalised as we believe ourselves to be, cannot speak and act in place of the people. That is why most of this chapter will be of a critical, destructive nature.[2]

Too often, up to now, and particularly in the case of contemporary physical sciences, demonstrations of the role of the dominant ideology have not been convincing enough; they were too biased towards the ideology they were to pinpoint, even in the very way the questions were posed. The problem in looking at ideology is to distinguish the reality that the name of the science (physics for instance) obscures, and hence the field of action for the critique. Implicitly, at least, this reality is limited to the discourse* of the science in question, essentially the discourse of research and discovery. Further, discourse is often related to a debased form (so-called scientific 'popularisation', even if done by scientists themselves) more than to the original form of the actual texts of research papers. Occasionally the epistemology of the subject is con-sidered, either as the work of 'philosophers of science' or of amateur-philosopher scientists; more rarely the pedagogy of the subject (the content of textbooks) is considered. In all these cases one deals essen-tially with the results of scientific activity (experimental 'facts', theoretical 'laws'); of its intellectual byproducts (philosophical com-mentaries); or of its material byproducts (technological applications). Yet by detaching the finished product of a scientific practice from the conditions of its production, the practice is so mutilated that it loses precisely those features most marked by the ideological environment, and provides only inadequate grounds for analysis. This notion of scientific activity bears the marks of the dominant ideology, particu-larly its refusal to ascribe all knowledge to the practice from which it proceeds; such an approach precludes effective criticism. I believe this

---

* [Editors' note]. Throughout this chapter we have rendered the French *discours* as *discourse*; though the English word is lacking something of the resonance of the French, where it refers to the whole set of statements, implicit or explicit, in a given area of social activity, and emphasises their common features.

methodology to be the relevant one for considering the attempts, made at other times and places, to distinguish a 'proletarian' science from the traditional 'bourgeois' science.

I wish to consider problems specific to physics, since it was in the fields of the theory of relativity and quantum mechanics in the Soviet Union that some of the sharpest attacks were developed against 'bourgeois' science. These attacks were centred on the current philosophical — and properly criticisable — exegeses of these theories, but they were none the less unable to escape the problematics of these same exegeses, which were presented as necessarily linked with the theories in question. Thus the criticisms did not achieve their purpose. The validity of the theories of contemporary physics, fully confirmed by a larger and larger collective practice, reduced to nothing the efforts of those who, too often, threw out the baby with the bath-water; starting with a justified criticism of the ideological exploitation (in the full meaning of the word) of the theories, they ended by claiming that the *content* of the theories was itself at fault. I deal again with these problems later in this chapter. For now, we may note that the failure of these attempts resulted in complete confusion, and the earlier dogmatism on ideological questions has too often given way to total laxity.

Today, nevertheless, the ideological climate in advanced capitalist countries both makes possible once more, and at the same time urgently demands, an examination of these questions. In order to avoid falling into the old trap, the critique can and must begin with the actual practice of contemporary science. In other words, ideological criticism cannot be limited to epistemological problems in the traditional sense, but should necessarily aim (even to confront these very epistemological problems) at the economic, political, sociological and even psychological implications of science, both in the form of its various practices at these several levels (different aspects of scientific production) and of the connection between these practices and other social events.

It is against this background that I should like to deal with some ideological aspects of contemporary physical sciences. The generality of the problems that I have just outlined makes it difficult to focus the criticism on questions specific to physics. This is the reason why several of the notions I shall formulate are, in my opinion, applicable to all or a large part of today's scientific activities. Yet, even from this global perspective, physics seems to me not simply one example chosen amongst others. It plays, in some respects, a canonical part, for through it various general aspects of modern science are expressed in a particularly concentrated form; the separation between questions of power (political, military, economic, and so on) and those of knowledge (epistemological, philosophical) is much weaker in physics than in many other disciplines – and this only increases the necessity of tackling the former before approaching the latter. It is, however, possible that certain of my generalisations are too broad, particularly as I shall not be able to discuss all of the contemporary physical sciences. For reasons which are subjective (because of my own professional practice) and at the same time objective (at least I hope so), the physics of fundamental particles and their interactions (high-energy physics) is given priority. Only a comparison with conclusions drawn from the analysis of other areas of science will indicate whether it is necessary to restrict such remarks, which I believe to be generally valid, to physics alone, or even to some of its specialisms.

IDEOLOGY, PHYSICS AND POLITICS

In accordance with the nature of the problems I have outlined above, I should like to start by considering the role of ideology in the relations that scientific activity – here, physics – has with other sections of social life; and – conversely – the role of these relations within the dominant ideology. I shall deal, essentially, with two types of relations: first, those which, up-stream of scientific production proper, condition it; and, second, certain of those which, down-stream of scientific production, mediate its repercussions.

Ideology and the Orientation of Physics

The problem is that of understanding the mechanisms which preside over the choice of priorities and directions of scientific development. According to what criteria do the private or public specialised agencies, of the ruling class decide to support – materially and morally – this or that discipline? Consider the example of high-energy physics. During the 1950s this branch underwent a massive development in the number of research workers and of experimental equipment (big accelerators) at an excessive cost. For instance, in the United States particle physics receives 27 per cent of the total federal and industrial research grants for fundamental physics (more than solid-state physics, which gets 24 per cent, though it is of a much higher technical interest; nuclear physics is third, and far behind, with 13 per cent). If we take into account equipment expenditure, more than half of the public funds available to physics is absorbed by particle physics. This is not surprising when, for instance, the National Accelerator Laboratory at Batavia, the biggest accelerator operating at present anywhere in the world, required $250 million for its construction and consumes $60 million per year in running costs. In Britain, government funds, distributed by the Science Research Council in 1967–8, amounted to £39 million for high-energy and nuclear physics, as against £20 million for astronomy and space research, and £16 million for all the other non-medical sciences.

Given the ever-growing esotericism in the field, where partial discoveries accumulate without reaching any form of definitive synthesis, one may raise the question of the reason for its importance. The sparse technological spin-off from the construction of big accelerators (improvement of electro-magnets with super-conductors is an example) is not a convincing justification; no doubt it could have been acquired with less cost by more direct technical research. But, to offset against this there are obvious economic mechanisms; as a place for regulatory public investment and a source of super-profits for specialised firms, the construction of big accelerators plays for capital a similar role to that of space rockets and satellites, partly replacing the arms race. This is why, when economic prospects contract, the credit tap is turned off, to the great distress of naive physicists who had believed in government declarations of faith in the fundamental interest of their research work.

Other necessities, more political (in the narrow meaning of the word), and, therefore, already ideological, exist; big accelerators make possible the setting up of multinational bodies which can be used, either as a model for states engaged in the elaboration of a common policy, or as an excuse for those launching into such common policies only with reserve – CERN in Geneva for Western Europe, Dubna in the Soviet Union for Eastern Europe, both explicitly perform this function. Nevertheless, these arguments are inadequate; other fields, such as astrophysics (large observatories) or plasma physics, could have played the same role, some with more foreseeable economic 'spin-off' (for instance research on thermonuclear fusion). There is a twofold explanation of the particular importance of high-energy physics related to its historical development.

(a) On the one hand, high-energy physics is seen as the present peak of 'fundamental' physics, the heiress of atomic physics and nuclear physics; it wishes to study the 'deepest structures of nature and its basic principles'. But today's prevalent understanding of the relations between the natural sciences is essentially linear, even if one recognises the existence of several possible 'levels' of research. These levels (for instance the atom, the nucleus, the fundamental particles) are also conceived as subject to one-dimensional order and one-way causality, so that events at the higher level can be reduced to those at the lower. It follows that particle physics appears to be the study of today's deepest accessible level, on the basis of which the study of other levels must be approached. One thus finds, at the very heart of the organisation of the various physical sciences, a well-defined epistemological hierarchisation, whose nature, however, is strongly ideological. Then, even if it were true that, for instance, 'nuclear physics can, in principle be explained by particle physics', the very phrase 'in principle' demonstrates the impossibility of a total reducibility and the relative autonomy of the field of study. In the same way, chemistry, even though re-established 'in principle' as atomic and molecular physics, has, none the less, continued as such and kept its own concepts and methods. Moreover, the hierarchy within physical sciences is very similar to the one which still orders the various natural sciences in their current form, that is physics, chemistry, biology, geology, and so on, according to Comte's old classification. In general, the ideological privileges of physics are thus explained as those of its most 'advanced' branch.

However, the study of nature cannot be reduced to the successive opening of a series of Russian dolls. Relations between scientific disciplines cannot be specified solely by the more or less profound character of the laws they establish. The mutual borrowing of concepts, of experimental techniques, connections with industrial applications, philosophical weight, and so on, all make a multi-dimensional view indispensable. As an example, if particle physics is *theoretically* and in principle the basis of nuclear, atomic and molecular physics and, therefore, of the physics of condensed matter, it is equally possible to claim that the latter is *in practice* (partly) the basis of particle physics, which experimentally makes great use of solid-state physics for its detectors, of super-conductivity for its magnets, and so on. The implicit, yet overwhelming belief in a unilateral hierarchy is part of the current conception of physical sciences, which, based on theoretical relations between these sciences, reflects the dominant ideology of bourgeois society; it explains, in part, the privilege conferred to particle physics.

(*b*) The massive development of particle physics originatedin the period after the Second World War; as a result of this it cannot historically be separated from the military applications of nuclear physics proper. The latter, which, until 1940, had remained a very advanced field, prestigious yet purely academic, suddenly showed itself as having considerable political importance. In addition, nuclear energy was put to work in practice by the very men who had discovered its theoretical potential; this is crucial by contrast with the nineteenth century situation concerning thermal or electrical energy, which were put to practical use by technicians and engineers, independently or almost independently of theoretical (often subsequent) research work. Everyone knows the story of the Manhattan Project and the participation of the most eminent theoreticians of the 1930s in the making of the first A-bomb. It is hardly surprising, therefore, that particle physics, natural offspring of post-war nuclear physics, has been given preferential treatment from political leaders, both as a token of gratitude for services rendered, and as an investment in possible new sources of power.

In fact, for a long time politicians were deceived or deceived themselves with regard to the possible practical applications of sub-nuclear physics, and physicists were very cautious not to let them into the truth too soon. Were not many sensational papers on 'neutron bombs' or even 'neutrino bombs' or thermo-nuclear fusion controlled by muons, and so on, published or allowed publication between 1950 and 1960. When it eventually became evident that particle physics would not have, in the foreseeable future, any profitable military or energetic applications, the mechanism was already set in motion and the indirect, non-specific economic factors mentioned above were strong enough to keep it going.[3]

This narrow utilitarian point of view also bears the mark of bourgeois ideology. The concept according to which all progress in scientific knowledge is necessarily followed by technical repercussions may have originated in a certain appraisal of the development of science between the seventeenth and nineteenth centuries, when it seems to have been closely connected with the forces of production. Perhaps this is also a delusion. Anyway, I should like to defend the idea – though it may seem paradoxical – that, at present, this is no longer the case. It is true that it is possible to list all the great technical innovations rooted in physics during the last century, from wireless telegraphy to transistors via nuclear energy, and show that the time gap separating a theoretical discovery from its practical application has constantly diminished, and deduce from it that 'science is becoming more and more a direct productive force' (as Marx had already written, although in a totally different context). Yet this argument can only pose serious problems of principle: must the discovery of the laser date back from its effective realisation by Townes in 1956 or from Einstein's theoretical work on stimulated emission in 1905. In other words, is it possible — and if so, how? — to single out the precise time of the discovery in a collective and continuous flux? An even more immediate aspect of this appears in those laboratory discoveries which find an outlet in industry. Perhaps they do find an outlet more and more rapidly, but proportionally they are less and less numerous. There is a real inflation of scientific 'production', characterised, for instance, by the staggering increase of scientific publications, the great mass of which essentially ensures the author's maintenance and the social status of his or her (ideological, in particular) function, and whose links with any kind of practical applications are increasingly tenuous. Justifying particle physics, 'the peak of research work', by its technical applications is therefore invalid. Many physicists — though it is true, in other fields than particle physics[4] — now admit this. The weak and defensive nature of this argument is today self-evident. Yet the argument seems indispensable to the panoply of supporters of 'the maintenance of the present situation', whether scientists in search of funds or ideologists defending the value of science. 'Every discovery eventually finds a use one day', 'one never knows, it might be useful after all', and so on and so forth. One can appreciate here that the weakening of the ideological motivations that used to support research in particle physics goes hand in hand with restrictions that are being applied to economic growth. Consequently, more and more sophisticated staging becomes necessary to the protagonists, such as the demonstrations organised at the international conference on particle physics, held in September 1973 at Aix-en-Provence. There, special sessions were set up for the public under the title 'Aix-Pop' (sic)* in order to 'popularise' particle physics. By transforming the subject into a show (exhibition of material, photos) it will perhaps be granted an even more mystifying role through this apparent accessibi-

---

*Lest these remarks might look somewhat elitist, let me make clear that I do not take here a patronising attitude towards the efforts of those who, often very earnestly, try to popularise modern science. In areas such as fundamental physics, however, due to their esoteric and spectacular character, popularisation might very well contribute to reinforcing the mystification of science, instead of destroying it. See, for instance, P. Roqueplo, 'Le partage du savoir (la fonction culturelle de la vulgarisation scientifique), (Paris: Le Seuil, 1975) and B. Jurdant, 'les problemes théoretiques de la vulgarisation scientifique', thesis (Strasbourg, 1974).

lity than the esotericism which has been the rule hitherto — all the more since the good intentions of the organisers were obvious, rightly pre-occupied as they were by social problems and the public image of 'their science'.

Physics and the Ideology of Expertise

If, as I have argued, the dominant ideology conditions in great measure the directions and rhythms of scientific development, the former is also influenced reciprocally by the latter. One of the most distinct aspects of the new form of this ideology, linked with the increasing impor-tance of science in 'advanced' capitalist societies, appears in the social function bestowed on scientists, or rather on their image. (I shall come back later to the manner in which the current picture of scientists is removed from reality.)

It is true that the *savant*** (since the term is still used in the mass media) has become the figure-head of the boat on which the masses are taken for a ride. Following the priest, the artist, the teacher, it is now the scientist who often best embodies the modern values of the domin-ant ideology: *expertise* and *competence*. Possessor of an 'objective' truth, that has been 'strictly demonstrated' and is 'politically neutral', the scientist vouches admirably for the attempts of the ruling class to hide oppression and exploitation behind would-be rational and technical necessities. Be it public transport, energy resources, tele-communications, the food industry, from Concorde to synthetic meat via nuclear power stations and colour television, scientific progress, shown as being ineluctable, is used to conceal the choices of a class and the ideological, political orientations of 'innovations'. When pay-slips are prepared by a computer and not by an accountant, when selection in schools is made on the basis of mathematics instead of Latin, when the inspection of tube tickets is no longer manual/visual but magnetic, the mechanism of the oppression of people by people is hidden behind the appearance of an oppression of people by things:

> The more we extract things from nature thanks to the organization of work, of great discoveries and inventions, the more we fall, it seems, into the insecurity of existence. It is not us who dominate things, as it appears, but things which dominate us. Now, this appearance holds good because some men, through things, dominate

---

*More than the English equivalent of 'scientific expert': a *savant* is a wise man!

other men. . . . If, as men, we want to enjoy our knowledge of nature, we have to add to our knowledge of nature our knowledge of human society.[5]

In order to institute new forms of oppression and exploitation (or old forms under new conditions) acceptable in the name of science, scientists are made the heralds of power. Of course this is only true for a few highly placed individuals of the scientific establishment whose prestige, resting on a traditionally elitist and meritocratic conception of science, is used to grant a universal validity to the ideology of expertise and competence. Put forward by an 'unquestioned specialist', every idea, even if it has nothing to do with the scientific speciality, is adorned with the power of absolute knowledge. This applies also to the traditional left, where Nobel Prize winners' signatures are highly valued on any appeal. Modern physics seems to play here a special role, in conformity with its own image as a peak discipline, and, therefore, one which is particularly invested with this ideology. Modern physics has, in recent times, supplied many of these more or less naive spokesmen of the dominant ideology. Television, for instance, deliberately exploits the image; in France, L. Leprince-Ringuet, ex-director of the High Energy Physics Laboratory at the College de France, thus became a true television star, even to the point of commenting on tennis matches and classical music! Many officers in organisations concerned with scientific affairs, and with more important duties than one would perhaps assume, are also physicists. This is particularly the case in international relations (just think of the recent journeys to China by eminent physicists).

One of the most remarkable examples of the role played by at least one of the branches of modern science, high-energy physics (again), is supplied by the list of the Jason Division Members. Attached to the U.S. Institute for Defense Analysis, this group of scientists,[6] acting as a Pentagon advisory body, has pondered over a whole series of questions, from electronic war in Vietnam, counter-insurrection in Thailand, military use of lasers, and anti-missile missiles, to techniques for repressive policy measures inside the United States. Now, this group was composed of a two-thirds majority of high-energy physicists, mainly theoreticians, obviously the least qualified a priori to discuss the technical applications of experimental discoveries. This can only be explained by a dual ideological process. On the one hand, the most 'fundamental' part of physics is thought of as the most important and the most difficult; its experts, therefore, are also the most intelligent and the most competent in *everything* — an opinion that they themselves evidently share. On the other hand, scientists can be given the illusion that it is 'they' who are going to manipulate the military, and not the reverse. Suspicious when they are asked to sell their material or their formulae — as they realise they would then lose control over them — scientists still believe that they will remain the masters of their brains while 'hiring' them. The record of the Jason Division, especially its role in Vietnam, clearly shows who uses whom. It has to be stressed that, if the Jason case is mentioned here, it is because it typifies the ideological role of science which is the object of our interest. But physics would not shield (even if, no doubt, it is one of its tasks) the existence of numerous other, less brilliant, less known, more technical, more efficient organs of co-operation between scientists and the military.

One should not forget to mention here W. Shockley, 1956 Nobel Prize winner in physics, co-inventor of the transistor, who has become an implacable propagandist in the United States of modern genetic racism in its most vulgar forms, again constantly referring to his so-called scientific competence in a field totally foreign to him.

Another interesting example, is that of the last French Nobel Prize winner in physics (1970), L. Néel, who has been appointed Chairman of the Highest Council for Nuclear Safety; in fact he is a specialist in . . . magnetism! But of course the committee (who appointed him) 'tries to stand above polemics [and] seems to be a body mainly intended for reassuring public opinion'.[7]

Generally speaking, the situation in this area, of the safety of nuclear power stations, is particularly interesting and revealing. On one side, in all Western countries, government experts, members of atomic energy commissions or of various bodies dealing with energy policy, have always spread reassuring, soothing words, even when, during the course of years, they themselves had continuously to reduce the radiation-dose limit believed to be acceptable, and even when new risks were detected (such as recently, after several years of running, the unexpected contraction of rods of fissile material, which subsequently caused American nuclear power stations to considerably decrease their energy output). On the other side, too often one finds other experts, liberals full of good intentions, who do not go beyond the apparently blind path of figures and theoretical debates. One may remember, as an example, the 'tournament' on the question of the danger of nuclear test fall-out between the 'good' Linus Pauling and the 'bad' Edward Teller, both armed with their Nobel Prizes, and relying on their scientific reputations on the question of the danger of nuclear test fall-out.*

Such technical discussions threaten to prevent all political intervention by the people by disguising the essential question of *power*. Who decides energy policy? Why the energy race? Why the disregard for local, controllable sources of energy in the name of 'profitability' which makes sense only for capital? These are questions that only can be discussed and solved by participation in a mass debate to which all technical evaluation should be subordinated.[8] In other words, radical competence, the expertise of the left, forms a dangerous trap of the dominant ideology. 'Be Red first, expert later' was a motto in China during the Cultural Revolution.

IDEOLOGY AND SOCIAL PRACTICE OF PHYSICS

Division of Labour and Hierarchy

So far I have spoken with little qualification of 'the scientists'. It is time to come back to the term, which beneath its apparent neutrality and universality, reflects the dominant ideology at the level of the sociology of scientific production.[9] In reality, the notion of a homogeneous scientific milieu, where the only difference between scientists is purely quantitative, is far removed from the truth. Certainly this notion has historical roots, and found its source in the position of scientists from about the seventeenth to the nineteenth century. Research work was then of an essentially individual nature, a kind of craft, both manual and intellectual. Scattered and relatively isolated, poorly institutionalised, scientists were divided into a hierarchy, basically built upon mutual acknowledgement of merits. But with the growth and the industrialisation of scientific institutions increasing steadily over the course of this century, the situation has changed radically. A division of work was imperative. Disciplines and even sub-disciplines became more and more compartmentalised; for example,

---

*Perhaps, I am somewhat unfair here to Pauling, and those people whose attitude was a very courageous and useful one in the Cold War atmosphere of the 1950s. This was a period where any dissent from the overwhelming dominant ideology was meaningful; the conditions were such that even individual liberal positions appeared as radical ones. Today, however, a much more collective consciousness is building up, and I wish mainly to point out that, in this *present* situation, elitism and expertise should be considered as most dangerous stumbling blocks for any radical critique.

theoretical high-energy physics only exists as a conglomerate of specialities – group theory, field theory, S-matrix weak interactions, and so on. The gap between theoreticians and experimentalists is often insuperable, for example in particle physics; ever more numerous and hierarchical teams became set up.

The picture of the master working with his devoted pupils and future successors and one or two faithful technicians, in an atmosphere of mutual help and respect, even if it reflects (albeit rose-tinted) a nineteenth-century reality, is definitely out now. Today, in research work, one finds bosses who spend most of their time in general administrative and political functions. Often they have lost real touch with the research, yet at conferences they still present the results of 'their' research workers. One finds 'confirmed research workers', who organise the work, distribute the tasks and watch them being executed. Often they are the only ones to have a clear idea of the current project. Finally, one finds research workers, with precarious status, indefinite position and salary, forced into submission in order to keep their chances of promotion. The real social hierarchy, that of power within and outside the institution, is not continuous at all, and only real rites of initiation (for example a thesis, an appointment as a director) enable one to climb up the ladder. There is no social agreement in the scientific milieu, but sharp conflicts of interest.

Moreover, the correlation between the hierarchy of power and that of competence is becoming weaker and weaker as the very notion of individual competence tends to lose all meaning. When experimental work becomes explicitly collective (scientific papers bear more and more signatures), when theoreticians themselves often do no more than progressively improve on a common idea, then individual recognition is largely due to factors outside scientific activity, such as a good public-relations system, easy access to experimental instruments, membership of a rich and prestigious institution. All these features are more or less conspicuous according to the discipline. In high-energy physics they are often at the level of caricature, but they are found elsewhere in various degrees in most fields of physics.

The ideological need for an image of 'the scientist' must be very strong for it to continue imposing itself despite reality, and many mechanisms operate to this end. One of the most effective, both in respect of the public at large and of the scientific milieu itself, is no doubt the system of the specialised attribution of scientific prizes, and particularly the most famous of them, the Nobel Prize. Founded during the last century, the Nobel Prize had then a certain meaning related to the actual state of scientific production, which was at that time, as described above, individual and artisan. The fact that it still operates with the same unchanged rules is evidence of its essentially ideological nature. To convince oneself, one has merely to consider the Nobel Prizes in physics for the last fifteen years. One first notices that particle physics recives the lion's share, collecting about half of the awards, whereas other fields like astrophysics, though fertile in recent discoveries (quasars, pulsars, and so forth), remain ignored. If prizes are shared by two or three laureates, it is very often in a rather arbitrary manner; for example, only three of the four founders of quantum electrodynamics were rewarded in 1965 (J. Schwinger, S. Tomonaga, R. Feynman), though the fourth man (F. Dyson) had played an essential part. On other occasions, only one individual receives the award (for example A. Kastler in 1966) though the contribution of his main assistant (J. Brossel) could not, in practice, be separated from his. It is a fact that, when evaluating true merits, the collective nature of a piece of work should be acknowledged; along with and beyond Kastler and Brossel, their whole laboratory at the École Normale Superieure should have been 'Nobelised'. In 1959, E. Segré and O. Chamberlain were awarded the Nobel Prize for the discovery of the anti-proton, a straightforward discovery if ever there was one! The only merit of the team which made it was to have had the first access to the particle accelerator (the Bevatron at Berkeley) which had been built for this purpose. Rivalry around the machine was intense, and it was in corridor discussions, and debates of the various committees in charge of organising the use of the accelerator, that most energy, imagination and intelligence was spent; thus a predictable Nobel Prize was won. This particular case has actually bounced back into the news again with the action brought against Segré and Chamberlain by another physicist, O. Piccioni, on the grounds that they had 'borrowed' an idea essential to the experiment without acknowledging the orginator, hence frustrating him of a share in that Nobel Prize. It is interesting to note that Piccioni claimed $125,000 compensation plus interest, that is ten times the amount of 'his' third of the prize, from which he considered himself to have been dispossessed. This gives a good idea of the various political and financial sequelae which the award of the Nobel Prize brings. We may also note the comments of C. Wiegand, co-author with T. Ypsilantis, Segré and Chamberlain (the original paper bore the four names) of the discovery of the anti-proton:

Wiegand, though understanding Piccioni's feelings and having participated fully himself with Ypsilantis in the experimental work on the
anti-proton, yet not in the acknowledgment which followed it, does
not want to take part in the controversy. . . . Wiegand insists on the
fact that few experiments, in physics as in other sciences, are truly
original. They are the results of many combined ideas. It just happened that the times was ripe for the discovery of the anti-proton.[10]

One could not describe better the diffuse collective nature of scientific
discoveries and the contingency of nominal rewards. Finally, Piccioni
explains his long seventeen years' silence on the grounds of pressure and
threats exerted against him by his two 'colleagues', till now more powerful than himself in the establishment.

Naturally, even in this case, no mention is made of other research
workers, engineers and technicians who all took part in the experiments.
In the case of the 1968 Nobel Prize, awarded to L. Alvarez for intensive
experimental work in particle physics, performed by the large team he
directed at Berkeley, we almost reach caricature. The group's papers
contained multiple signatures; taken at random among their plentiful
production, an article published in 1964 has sixteen signatures and ends
with the traditional and paternalistic and revealing thanks:

> we are grateful to the technicians of the 72 inch bubble chamber and
> of the Bevatron for their competence and their patience. We also
> thank our team of scanners and film measurers for their continued
> efforts without which this experiment would not have been
> possible.[11]

But, of course, only the boss of the team cashed the Nobel cheque and
benefited from the prestige of the prize. The situation is not fundamentally different, despite appearances, among theoreticians. Thus, Murray
Gell-Mann received the prize in 1969 for his many contributions to the
theory of fundamental particles. But most of these contributions are
referred to using his and another name, often that of an independent
author of the same discovery, for example the formulae of Gell-Mann
and Nishijima, or of Gell-Mann and Okubo, the unitary symmetry of
Gell-Mann and Neeman, the quarks of Gell-Mann and Zweig, and so
on.

One can verify here one of the conclusions of a very interesting work
on the sociology of the Nobel Prize:[12] based on work of equal scientific
merit, the probability of getting the Prize largely depends upon reputation already acquired by the individual as attached to a celebrated

university, or on being a pupil of a previous Nobel Prize winner (there are real hereditary Nobel dynasties), or on being an individual of consequence in scientific politics. Real campaigns take place, organised by this or that government (France, to mention but one example) or lobby (technocrats of arms control) for the award of the Nobel Prize to certain individuals.

What is said above about the Nobel Prize is valid for the whole system of social rewards available to scientists: progression in the hierarchy; range of awards at all levels and on every scale; appointments to prestigious offices (from Academics to Committees of Wise Men); and access to the public tribunes of the written or spoken press. In all cases, it is less the strictly scientific competence of the individual – as I have argued, this has tended to lose its meaning – which is rewarded, than an ability to operate within the scientific institution and to make it work. The existence of an important network of acquaintances in the milieu, access to decision-making bodies, skill in the struggle for power and recognition, accommodating with competitiveness, complicity with colleagues, these are the factors of success. This system of rewards plays its role very well; it channels energies into the roads of conformity with the competitive, elitist model, and sustains a hierarchy perfectly adapted to its political and ideological role. The difference between real criteria in the ascent of this hierarchy, and ostensible motivations, only reflects the distance between the effective functioning of the scientific institution and its image – the precise distance that the dominant ideology maintains.

Division of Scientific Practice

In order to unmask this ideology, it was necessary to analyse, behind 'the scientists', various categories and their contradictions; similarly, it is now appropriate to analyse, in more detail, what has so far been called 'scientific production'. This term has been purposely used where, normally, one would speak of 'scientific research'; this is because exactly as the image of the *savant* relates to a small and old-fashioned stratum of the scientific world, the image of research corresponds to only a part of scientific activity. Even taken in its strictest meaning, that is the practice of scientists belonging to public institutions (universities, research centres), this activity has at least three aspects.

(a) *(Fundamental) research.* This is the one which, roughly, corresponds to the current image. It is the work involved in discovering laws,

properties and still unknown phenomena. However, developing theories and performing really new experiments is a relatively rare activity and represents only a very small proportion of scientific production. A very limited number of scientific departments, and within these departments only a small proportion of research workers, participate in this explicit investigation of the 'new'. If one refuses to discount as useless a vast proportion of current scientific work, if one does not want to fall into an ultra-elitist conception of science, then one has to recognise the existence and the necessity of what I should like to call (fundamental) development.

(b) (Fundamental) development. Short of better words, I am borrowing here currently used terminology: 'research and development' for 'applied science' (the distinction of the latter from 'fundamental science' being as obscure epistemologically as decisive institutionally). On the fundamental side, therefore, I shall call 'development' that large proportion of current activity which consists of applying already established methods or results to solving various problems, for instance the use of new theories to explain known experimental results and recent experimental techniques to test hypothetical predictions. Activity here consists less in searching than in finding. I mean that in this type of work, there is generally no doubt as to the possibility of success; the theory or the apparatus utilised *does* work and *must* yield results. Nowadays, whole sections of physics depend almost exclusively on the second type of practice. This is the case for acoustics, a great proportion of optics, the mechanics of fluids, and so on. But even in what are considered more 'modern' fields, following the perfection in the 1920s of quantum theory (research phase), advances in atomic and molecular physics (from spectroscopy to quantum chemistry) nowadays depends on 'development'. Other branches, such as mathematical physics, have, by their very nature, an almost exclusively developmental character.

The distinction established here between research and development coincides only partially with that of Kuhn's between 'revolutionary science' and 'normal science'.[13] or that, sometimes used, between 'extensive' and 'intensive' research. I shall come back to the connections between these various categories on another occasion (see, however, pp. 155 – 61).

I still, however, have to take into account the role of teaching.

(c) Teaching. Since for its maintenance and pursuit, scientific activity relies on the training of specialised (wo)-man power and the moral

support of a certain proportion of society, it is clear that the diffusion of scientific knowledge is included in general scientific practice. The large proportion of scientists belonging to teaching bodies (universities) is the institutional expression of this.

The growing gap between these three types of practice is another aspect of the increasingly extensive division of labour. It is particularly clear in physics, especially in its most industrialised sections. Again, high-energy physics offers a good example. It is probably the field where the proportion of scientists working full time in research laboratories, with no teaching duties, is the highest. This can obviously be explained by the necessity of making profitable use of expensive equipment, whether particle accelerators or computers. Even between laboratories, there is necessarily a distinction between those (even if they are a minority) with the material means, in terms either of the power of the machines, or the quality of the libraries, or even the network of scientific relations, which make possible *avant garde* research (in the strict sense defined above), and those poorer, older, more isolated laboratories that have to confine themselves to simple development work. The fact that this discrimination is implicitly confounded with a value judgement (the 'better' a laboratory, the more advanced is the fundamental research) is an index of the ideological effect we are interested in. Once more, in our present society, a non-hierarchical categorisation is not possible. If, in the way science is presented to the general public, and even in the image that scientists themselves form of it, there is confusion between the practice of science and that of research (in the strict sense), this situation expresses the privileges effectively conferred on the latter compared with development and teaching practice.

If one wants examples of these privileges, one may look, once more, at the list of Nobel Prize winners. The Nobel Prizes are generally conferred for research work and not to contributions to the development of physics, however important. There are a very few exceptions, such as, for instance, the Prize awarded in 1912 to an engineer, G. Dalén, 'for his discovery of automatic regulators which, combined with gas accumulators, are used for the lighting of lighthouses and for light buoys'; and also the 1973 Prize won by L. Esaki, I. Giaever and B. Josephson for their work on the 'tunnel effect' (theoretically known for fifty years) and on semi- and super-conductors. In addition, one may mention G. Herzberg, who, for the important work of his group in molecular physics (which depends, according to the definition employed here, essentially on development), in 1971 received the Nobel Prize for . . . chemistry! The Prize, according to Nobel himself, should be awarded to the individual who has made the most important discovery or *invention*; the criteria for 'important' were not defined. If, finally, the importance has almost always been appreciated in relation to peak research, it is not very surprising when we consider that development is a much more collective work, in which it is even more difficult to distinguish each individual contribution. Another example, illustrating this time the practice in teaching, is given by the criteria for appointing -- at least in France -- university professors and lecturers; normally only their qualities as research workers are generally taken into account, and very rarely their teaching skills -- often very poor as many of them had never taught before they were appointed!

The hierarchy of values, set up as it is between various aspects of scientific practice (I shall describe later some of the consequences of this), obviously finds its source in the general elitism of the dominant ideology. But there are also more specific causes, linked to the conquering spirit of the formerly triumphant bourgeoisie. Humanity's desire for knowledge has not always taken the form of the will to push continuously out the boundaries of knowledge, as it was during the period of the great geographical explorations of the nineteenth century. We know well how that knowledge, when brought back, helped the power of imperialism. The privilege conferred today to an extensive and more and more esoteric knowledge, as opposed to an intensive knowledge, would also point to a certain view of natural sciences as one of the rare fields of expansion, apparently indefinite, still open to the bourgeoisie's determination to rule. If the domination of humans over things (and where to find things, more purely as things, than outside the social sphere and even outside life itself, that is in the world of *physical* sciences?) serves to mask the domination of humans over humans, one can then see that the growing revolt against the latter demands the increase of the former. And the more the forms are incomprehensible to the majority of people, the better scientific research is able to play its dual role as an entertainer for diversion and as a guarantor of the hierarchy.

A magnificent example of this dual role is provided by the image of Einstein in the properly mythical dimension it has acquired at the popular level: author of a theory 'that about only a dozen people were able to understand' (expertise), whereas relativity is now currently taught during the first year at university; and, at the same time, sorcerer's apprentice, reputedly responsible for the atomic mushroom of Hiroshima (picture of terror). It is the usual irony of history if the last physicist of this bygone era, when science wanted to be artisan, disinterested, humanistic, has become the opposite: a symbol of the industrialisation, exploitation and oppression which are now clearly associated with science. One understands, therefore, why science is generally presented as an exciting activity of discovery performed by a few over-gifted geniuses, whereas it is in reality mostly a day-to-day, sometimes routine work, performed by wage-paid workers.

IDEOLOGY AND EPISTEMOLOGY: IDEOLOGICAL EXPLOITATION; EPISTEMOLOGICAL RE-CASTING; PEDAGOGICAL LAG

The triple division of scientific activity, into the hierarchisation of individual functions, the specialisation of disciplines and the division of labour in practice, which is particularly advanced in contemporary physics, has considerable internal consequences. The privilege conferred to research (narrow sense) leads to an excessive importance being given to fundamental discoveries, exceptional breakthroughs and qualitative advances, which represent the most unforeseeable and irregular moments in the history of scientific progress. The activity of development, nevertheless, plays an essential role in freeing the new-found knowledge from the mud of the old formulations (this new knowledge is always, at birth, prisoner of the old formulations and at the same time the proof of the need to leave them behind). It is this development which makes it possible to extend, little by little, the area over which concepts are valid, and the functioning of experimental instruments, to the limits demanded by new fundamental progress. Consequently, if scientific knowledge advances by means of qualitative leaps, it also remains the object, between these leaps, of a constant process of re-casting, essentially under the influence of development and teaching practice.

A good example can be found in the history of classical electromagnetism, which, after a long and rich pre-history, mainly between the seventeenth and the nineteenth centuries, truly began with Maxwell's formulations of his famous equations. At the time, these were included within a mechanistic context which would eventually be demonstrated as being invalid. Though the equations of the theory, within today's limits of recognised validity, still hold good (yet it would be interesting to analyse the progressive condensation of their symbolic expression leading to simultaneous simplification and enrichment), the concepts underlying them have none the less changed profoundly. Following a long period of work (experimental and technical; from Hertz to the development of telecommunications; theoretical, from relativity to general field theory), it is the concept of the electromagnetic field which has progressively developed over the last decade.

At a period when the various aspects of the division of labour — already discussed — were not so well defined as they are today, the process of re-casting was largely spontaneous and taken for granted. It is no longer so; today one witnesses a paradoxical situation where the most modern fields of physics are, in one way, the most archaic, and their necessary re-casting suffers from a long delay. The most striking example of this situation is provided by quantum physics. Several decades after its foundation, it is still not treated with all the seriousness that its advanced age, its established reputation and its unquestioned efficacy (materialised in the transistor and the laser for instance) should deserve.

At present, everyday in the work of thousands of research workers, its practical application has enabled the emergence of a new 'sense of physics', of a specific theoretical intuition, and of a proper heuristic approach. In addition, long theoretical work on its formal structure makes it possible to free the architecture, to make its key concepts explicit and to leave behind the necessary confusion of the phases of conception and growth.

Nevertheless, this rooting in practice and reforming in theory remains for the most part latent and implicit and does not generally appear in pedagogical, epistemological or philosophical discourses on quantum physics. Thus, these discourses are encumbered with obsolete terms, statements and pseudo-concepts; for example, uncertainty relations, quantum jumps, subjectivity of the measurement with respect to the observer and observables, wave/particle duality, complementarity, disturbance of the system by the measuring instrument, and so on. All this terminology describes less the real structure (the actuality) of quantum theory and its relation to experience, than a certain conceptual, temporary scaffolding which, as useful and unavoidable as it might have been in the past, today only hides (to say the least) or (even) restrains ongoing construction.[14] The time for the necessary re-casting having come, an indefinite wait will only exacerbate the already visible signs of sclerosis and deterioration.

In maintaining the proper ideological distance between an implicit practice and an explicit discourse, an essential role has been played by the dominant philosophical current within modern physics, namely positivism and its several variants (operationalism, nominalism, and so on), as it manifests itself, for example, in the epistemological concern of many of the founders of quantum theory within the Copenhagen school. This current certainly initially played an important positive role in freeing a whole generation of physicists from the yoke imposed by the mechanistic rationalism which was previously dominant. By 'relativising' theoretical concepts and stressing the role of experimental procedures, positivism has made possible the adoption of radically new ideas, as quantum physics demanded them. But from its beginning, quantum physics led physicists to transgress most of their new philosophical dogmas, which now merely sterilise and immobilise quantum epistemology. Think, for instance, of the debate on the quantum theory of measurement, which had been declared closed or even void of sense by the orthodox positivist view as early as the 1930s. Nevertheless, this debate has lately seen new and important developments which open the path for solving many paradoxes and problems attached to the so-called 'failure of determinism' in quantum mechanics. A sign of this philosophical domination, which highlights its clearly ideological character, is its almost exclusive imposition by theoreticians (the elite, of course!). Among the authors of the abundant literature devoted to the questions of the basis and the interpretation of quantum theory, such as Bohr, Heisenberg, Born, Dirac, Von Neumann, Wigner, de Broglie, Einstein, Schrödinger, Langevin, where does one find the name of an experimentalist? One apparent exception might be Bridgeman, the high priest of operationalism. But in fact he confirms the rule, since his own field of experimentation was high-pressure physics, which has no direct connection with quantum physics about which he set out to philosophise. Hence one should not be surprised if, in the whole of this literature, the references to experimental practice – even as numerous and as continual as they are – remain purely abstract.

As for specialists in the philosophy of science, they are also now the victims of the fragmentation of knowledge. They have less and less access to true scientific knowledge as experienced, and generally they must be content with the vulgarised interpretations that some scientists are willing to give them. This knowledge, in the course of the process of vulgarisation, loses its active quality and becomes purely passive. No longer an analytical tool, it instead carries the weight of ideological and philosophical positions which are not so strongly and not so perniciously expressed in the original and 'proper' scientific papers. These positions are generally claimed by the scientists to be *necessarily* linked with the scientific truth of the theories and methods under discussion (this is the very mark of positivism); but, in fact, they only are re-actualised echoes of external and earlier philosophical theories and methods. Yet, short of a proper scientific practice, those who receive the theories and methods at second hand are not able to separate the core of rational knowledge from the ideological shell. The latter then grows gradually thicker with successive editions, whether secondary education texts, or, even worse, popularised versions for the 'public at large'.

In this way, major advances in modern physics, especially in relativity and quantum mechanics, have paradoxically fed an intensely irrational current. One knows the popular expression for scepticism and unconcern: 'everything is relative . . . as Einstein said' (and this is not so harmless as one would believe). At a seemingly more elaborate level, the mad attempts of Bergson to criticise and reinstate the theory of relativity within his own philosophy, even if they took place fifty years ago, still give evidence of a serious crisis in the relations between science and philosophy. The case of quantum physics is even more obvious. The alleged 'crisis of determinism', which, quantum physics brought about (in fact, it was a modification of the forms of physical causality), opened the gates to a flood of philosophical, ideological and even political lucubrations. Heisenberg's 'uncertainty principle' (very wrongly named too!)[15] was generalised out of any proportion and combined with the vague notion of 'complentarity', laxly applied and beyond its field of validity. It was thus used to assert the impossibility of understanding both the physico-chemical and biological properties of living matter at the same time:

> each time physical and chemical laws could be applied to living organisms, they found their confirmation there, therefore it appears there is no more roon for any 'vital force' alien to the laws of physics. But this is precisely the argument which has lost most of its weight because of the theory of physics.[16]

Many of Bohr's papers expressed the same view. Yet, twenty years later, the prodigious development of molecular biology, partly due to the role of quantum chemistry, refuted these sceptical views. Thus it is rather amusing to see the same 'uncertainty principle' called to the rescue of an illustrious biologist in his own ideological reflections today: 'Finally, there exists on a microscopic scale, an even more radical source of uncertainty, rooted in the quantum structure of matter itself. Now a mutation is in itself a microscopic quantum event, to which, consequently, the uncertainty principle can be applied.'[17] It is true that a few lines further one can read: 'one must underline that, even if the uncertainty principle had to be forsaken one day, it would nonetheless remain, etc.'.

How can one admit more clearly that quantum physics has intrinsically nothing to do in the discussion and is an authority cited merely to buttress an argument? Even more of a caricature was the explicitly political use of quantum physics, quoted against Marxism:

> Eastern (sic!) Marxism teaches that communist economy is a historical necessity, and from this conviction stems its fanaticism. . . . Physics has now developed the statistical interpretation of the laws of nature which correspond better to reality; from this new viewpoint, the belief of the communists in the inevitable realization of Marxist predictions seems grotesque.[18]

It would be easy to quote numbers of similar examples. The 'free will' of the electron was invoked to prove that of humans. Under the name of 'complementarity', general laws of social evolution were even opposed to those of individual behaviour. In brief, one has witnessed a true *ideological exploitation* of modern physics. It must be added that the philosophical idealism, which made this procedure possible, was shared by the very people who proposed to fight it on behalf of materialism. A complete rejection of relativity and quantum physics because they were 'idealistic bourgeois theories' — as was the case in the Soviet Union during the years 1940 – 50 — partook in the main of the same conception of science. This conception was at the same time too restrictive (limited to the results of strict practical research work) and too large (linked to the philosophical and political discourse which claimed to be built upon it). The proper ideological space in which to draw the necessary demarcation line was thus concealed.

If today that kind of argument has lost its virulence, it is precisely because the practical rooting of 'modern physics', both in the work of a large scientific collective, and in the many technical products created thanks to this 'modern physics', has made it far less convincing. How, today can one preach idealism, subjectivism or irrationalism in the name of a science which, from transistor radio to laser bomb, demonstrates its efficacy and rationality to the contrary in the hands of those who dominate it? It is upon biology (molecular biology in particular) on the one hand, and upon social ('mathematised and structuralised') sciences, linguistics for example, on the other, that new ideological arguments, delivered under the name of science, tend to be built today. This is why previous considerations about the ideological exploitation of the content of modern physics are inclined to lose some of their importance, and have occupied, so far, only a relatively limited place in this discussion, whereas, even fifteen years ago, they would have constituted the essential part of a text devoted to these questions.

However, if the philosophical and epistemological crisis of modern physics has lost some of its importance outside its own sphere of activity, it is different inside, where it continues to have serious repercussions in teaching. Devaluation in teaching practice, and delay in re-casting, each leaning on and perpetuating the other, are here fully revealed. As far as the theories of relativity or quantum physics are concerned, the last fifty years have hardly witnessed any major evolution in their mode of presentation. Most handbooks are surprisingly similar, repeating indefinitely the same schemes of inner organisation. As a general rule, a historical or rather chronological introduction — of dubious accuracy — is followed by some philosophical reflections in which traditional dogmas are enunciated under a much more schematic and poorer form than that of their creators.

Having fulfilled this first task, the author then approaches the 'strictly scientific' content of the book. It consists, in general, of purely theoretical, exaggeratedly formalistic accounts, from which references to real experiments steadily vanish. Not a single impression is left of the real procedures of scientific activity, of the dialectic between theory and practice, heuristic models and formalism, axioms and history. Modern physics appears as a collection of mathematical formulae, whose only justification is that 'they work'. Moreover, the 'examples' used to 'concretise' the knowledge are often totally unreal, and actually have the effect of making it even more abstract. Such is the case when the explanation of special relativity is based on the consideration of the entirely fictitious spatial and temporal behaviour of clocks and trains (today sometimes one speaks of rockets . . . it sounds better . . . but it is as stupid!). This kind of science fiction (which is not even funny) is the more dangerous as it erases the existence of a large experimental practice, where the theory of relativity is embodied in the study of high-energy particles, involving hundreds of scientific workers, thousands of tons of steel and millions of dollars. In this field, things are beginning to change, but after a long delay.

In quantum physics, the situation is even more serious. Introductory courses are often limited in fact to the study of the resolution of a particular partial derivative equation (Schrödinger's equation), without stating its real significance. An excessive and often archaic formalisation (wave-mechanics is sometimes privileged compared with other more general forms of quantum theory) hinders any systematic conceptual work. Only very recently, at an introductory level, a certain breakaway from the rigid tradition has occurred, and new, truly modern, textbooks have been produced, for example those of Feynman or Wichmann,[19] though these books show the necessity of going still further. Characteristically the use of these books has met strong resistance on the part of teachers, under the pretext of excessive difficulty; yet they are technically (mathematically) *much* simpler than the traditional ones. It is, then, the effort at conceptualisation, which such books have the great merit of demanding, which is refused.

This teaching situation, even if it appears unhealthy and harmful with regard to the simple aims of training and teaching (transmission of knowledge), is however in perfect ideological harmony with the general context of modern physics. A closed arduous, forbidding eduction, which stresses technical manipulation more than conceptual understanding, in which neither past difficulties nor future problems in the search for knowledge appear, perfectly fulfils two essential roles: to promote hierarchisation and the 'elite' spirit on behalf of a science shown as being intrinsically difficult, to be within the reach of only a few privileged individuals; and to impose a purely operational technical concept of knowledge, far from a true conceptual understanding, which would necessarily be critical and thus would reveal the limits of this knowledge. This is why discussions about educational problems take on the form of ideological struggle. It is also why, because of the essentially political nature of the resistance to change in this field, no reformist illusions should be entertained as to the possibility of any major successes, as long as such a struggle only relies on the internal critique of scientific workers and teachers, remaining within the framework of an unchanged technical and social division of labour.

The Frontiers of Science

The dominant ideology therefore plays a considerable role in adjusting the various forms of practice which contribute to the development of each sphere of science (here, physics), by way of control over their relative emphasis and relative advance. However, even more than within the various fields of scientific knowledge, it is in the tracing of their frontiers that the influence of ideology is felt at the epistemological level. Each scientific discipline needs a system of representation, norms and values which give it definition and set its limits. The implicit consensus of all scientists in the same discipline over this truly specific ideology is essential for their activity to take place, since this is the price they must pay to work within a common understanding of problems. The type of questions which are permissible, the approved procedures of demonstration, the accepted methods of work, the authorised forms of communication, are thus fixed.[20] The most obvious purpose of these partial ideologies is to permit the specific constitution of each particular science, by establishing its place and boundaries within the over-all field of scientific knowledge.

So 'physics', like other sciences, cannot be defined once for all in an abstract and definitive way by referring, for example, to the physical 'method', even less to the aims of its study. On the contrary, it is precisely scientific work itself, starting from the reality being studied, which defines its own proper objectives. The best proof of these ideological results is given by the historic development, the modifications of the borderlines (and, on occasion, the change of name) of each discipline.

Thus the fact that between the eighteenth and nineteenth centuries physics replaced 'natural philosophy' could not fail to produce a serious modification in point of view. And, even though it may be justified, there is a real abuse of language in calling Galileo, Kepler and Newton 'physicists'. One also knows how long and complex was the pre-history of electricity and magnetism. Before being assigned to 'physics', at the end of the eighteenth and the beginning of the nineteenth century, they were, for a long time, related to conceptions much closer to the life sciences. It would be very interesting to study, in detail, the borderlines between contemporary physics and the other sciences, and the territories of disciplines with mixed status. For instance, the periodically renewed and still unsuccessful attempts to constitute a science of 'biophysics' doubtless reflect the existence in physics of an ideological current at the same time 'reductionistic' and 'imperialistic', a refusal to recognise the possibility of autonomous yet rigorous biological sciences. Yet molecular biology, for example, has acquired characteristics which are unquestionably scientific, with a well-defined field of work and proper concepts and methods. It has really become 'a science', whereas the word 'biophysics' does not seem to cover more than a heterogeneous collection of biological problems to which the experimental theories and methods of physics can be applied, problems which do not seem to be the most numerous nor the most fundamental.

Conversely, it may seem curious that the development of quantum mechanics, and its lightning success in molecular physics, has not brought a great part of theoretical chemistry back into the bosom of physics. Though many statements of principle from the golden age of quantum physics during the 1930s proclaimed that chemistry could in principle be reduced to physics, it was essentially as quantum chemistry that this has developed. I discuss here the position of this particular branch of chemistry, which, from all theoretical and experimental points of view, is in perfect continuity with what is called 'molecular *physics*'; it is not a question of the (legitimate) existence of chemistry in general. One can see in the existence of quantum chemistry, it seems to me, a true phenomenon of rejection on the part of physics, based on the devaluing of an area of scientific work essentially devoted to *development*. We have, therefore, exact symmetry with the previous case, where the attempt at incorporating part of the life sciences into physics, clearly corresponded with the (real) character of fundamental research as defined above. The combination of two hierarchies, that of different sciences and that of differing scientific practices, leads once more to a confirmation of the elitism which impregnates contemporary physics.

However, the problem of the real definition of each science is not only — and is indeed less and less — epistemological, even taking into account the underlying ideology. The growing institutionalisation of science, and sociological sluggishness, make the question somehow trivial: would not today's physics simply be the activity of physicists, and they themselves be defined through their belonging to a laboratory or institute of physics? As long as one remains within the scientific establishment, the answer seems to be clearly yes. Then, perhaps it is necessary to ask a question which is in the long run more important than that of frontiers between the various sciences: where are the lines which, in a given field of knowledge, settle the boundaries of what can properly be called 'scientific'? Or, more precisely, how can we recognise a science from other forms of knowledge and action related to the same reality?

Before posing the problem as a whole, it would be as well to consider the usual distinctions between 'fundamental' and 'applied' science and between 'applied science' and 'technology'. Today, these distinctions are at stake in a conflict between, on the one hand, the dominant ideology, in the dichotomy it has established between theoretical and practical, intellectual and manual, elite and mass, always with the privilege attached to the first term of these pairs, and, on the other hand, the economic and political needs of capitalism, which tends more and more to subordinate knowledge to power, research to production, science to profit. I have already shown that the particular status of particle physics was the result of this interplay of contradictions. It is also in this perspective that one can understand why the *corpus* of a science does not develop in a cumulative way, but — independently even of the re-casting of the existing or new parts — sees whole areas breaking away from it. Disciplines such as acoustics, the mechanics of fluids, elasticity, and even classical thermodynamics and the resistance of materials, which were at the zenith of physics in the nineteenth century, are increasingly slipping out of the field of 'contemporary' physics. They are practically no longer taught at university, and in spite (or because) of their considerable importance in production, on account of their lack of ideological prestige (as they are not 'top' subjects, development has taken precedence over research), they have very little place in the popularisation and public image of physics in general. Physics — in its strict, maybe academic sense — the physics which is explicitly meant in such expressions as 'contemporary physics', progresses more through removal and exclusion than through continuous expansion. Today's physicists, apart from their own exaggerated specialisations, and considered collectively, do not know *more* than last century's physicists, they know *different*.

Based on this problem is that of the frontiers, not to say barriers, between any one science, with its apparent ineluctable esotericism, and familiar knowledge linked with current everyday practice in the same domain of (natural and/or social) reality. A considerable mass of practical knowledge still conceived not so long ago as 'scientific' is at present directly incorporated into the industrial — sometimes even artisan — development of production. Here lies the core of the problem; the theoretical content of this science is devalued or forgotten to the benefit of a purely technical conception. P. Langevin once described how the notion of electric potential, included in the final-year programme of his French Lycée, represented, at the time of his education, the acme of theoretical abstraction and of difficulty. The daily use of electricity has concretised this concept today in the familiar '110' or '220' volts of electric bulbs or domestic applicances, in the discharge children get in their fingers when they touch objects around them, in cattle behind electrified fences – and in the electric chair which was used to kill Julius and Ethel Rosenberg. Yet this knowledge remains partial and empirical. Here, the division between theory and practice reaches rock-bottom. For years, French children were taught the laws of Ohm, Kirchoff, Joule, and the theory of simple circuits, without knowing how to replace a fuse or to repair a two-way switch. The situation is perhaps even more preposterous in classical mechanics, whose concepts (speed, force, power) and technical applications (from wrist watches to machine-tools) are at the same time omnipresent and perfectly dissimulated under daily usage. One reaches, therefore, the paradoxical situation in which the more a science is involved in daily production, or even more simply in everyday life, the more it loses its 'scientific' nature. How many would think, while discussing 'contemporary' physics, to include in it, not just the electrical techniques of domestic circuits and the telephone and the mechanics of an automobile, but also the elementary electronics of radio and television, the fluid dynamics of plumbing, the physical chemistry of 'amateur' photography, and so on?

One can conceive, however, all the implications, first educational, then ideological, that teaching and training based on this kind of daily practice would have. But knowledge is truly reified, and once it has got out of the specialists' hands, it loses its 'scientificity', in the sense that every theoretical assimilation and therefore criticism is excluded to give room to mere empirical manipulation. The consequences are tremendous and contribute more than a little to the support of the ideology of expertise and competence.

Today, there is an increasing demand for 'specialists', and not just in modern technical-scientific achievements like nuclear power stations, Concorde and the moon race. To repair the smallest domestic appliance, a car break-down, or a gas leak, one relies more and more on a 'specialist', even if he is less prestigious – though it would be very interesting to analyse the differences in status between the repair professionals, the nice-man-of-all-trades-round-the-corner, and the 'after-sales service' provided by big firms in a very impersonal and 'technified' form. The break which severs institutional science from the social practice that once originated from it is certainly an imperative of our class society. The fact that it has been for so long implicitly considered ineluctable, and as an element of science as such, only reflects the degree of ideological impregnation of the concept of 'science' itself.

If one poses the problem of the break between science (*strictu senso*) and post-scientific practice, in which science is at the same time embodied and excluded, one must pose also the question of the discontinuity between science and certain *pre-scientific* practices. In fact, large areas of human activity exist which have permitted the long-standing accumulation of empirical knowledge in this or that sphere of reality, a great deal of which has remained at the margins of scientific development, which could – or should – have incorporated it. Indeed, it is from this knowledge that modern science, especially physics, developed, particularly around the time of Galileo:

> *Salvati:* The constant activity which you Venetians display in your famous arsenal suggests to the studious mind a large field for investigation, especially that part of the work which involves mechanics; for in this department all types of instruments and machines are constantly being constructed by many artisans, among whom there must be some who, partly by inherited experience and partly by their own observations, have become highly expert and clever in explanation.

> *Sagredo:* You are quite right. Indeed, I myself, being curious by nature, frequently visit this place for the mere pleasure of observing the work of those who, on account of their superiority over other artisans, we call 'first rank men.' Conference with them has often helped me in the investigation of certain effects including not only those which are striking, but also those which are recondite and almost incredible.[21]

But, once autonomous, science has progressively moved away from its origins and ceased to draw its fruitfulness and inspiration from the mass of popular knowledge. I wish to be clearly understood; I do not intend to make a fetish of the latter, which is generally a mixture of real but empirical knowledge and of pure superstition. Yet the point is that scientific rationality could make the sorting out possible; that is reject the latter and use the former. This type of attitude is well exemplified by the Chinese approach to 'contemporary' medicine which uses modern scientific methods, but, at the same time, instead of excluding traditional medicine, tries to assimilate it.

In the case of physical sciences, a good example is meteorology. The large number of sayings and proverbs on the subject shows the existence of a popular knowledge. There might be evident nonsense in many of them, but it is clear that, besides the nonsense, there is some real and useful information (anybody who has lived in the country for a while will know what I mean). A whole meteorology of micro-climates, in particular, could be derived from this popular knowledge, while, at the same time intervening to eliminate the obvious widespread illusions that also exist. Similarly, who has ever so far considered the relations between modern physical chemistry and culinary art;[22] yet, many recipes are, empirically of course, built on this or that process of the breaking-down of proteins, the dissolution of lipids, and so on. Are there not here interesting scientific possibilities which would both simplify and help understand cooking, while perhaps posing basic problems of physical chemistry?

My final example is a delicate one, that of water divining. In the country the practice still exists of looking for water with the help of water diviners and their twigs. Is it merely superstition? The apparent regular success of the method and its persistence seems to prove that it is not. Then, is there more than a simple empirical and perhaps unconscious knowledge of local topography and geology? Some people think so. Y. Rocard, ex-director of the École Normale Supérieure laboratories, also responsible for the initial research work of the French nuclear military programme, and hence a scientifically and socially orthodox physicist, but endowed with a particularly original way of thinking (a combination which is tending to disappear), proposed a physical theory of water divining,[23] and it is disturbing to realise that it was met with virtually total silence. Yet we have here a phenomenon which without doubt interests many more people than such-and-such a particular detail of the effective section of proton – proton collision at high energy. Refutation or ratification of theories of water divining should deserve more attention. But, once more, the gap between the science of specialists and the knowledge (or pseudo-knowledge) of the people is confirmed.

What this is the status of a 'truth', which, at present, can be evident to only a minority? What else can one conclude about the persistence, if not the intensification, of beliefs in astrology, extrasensory perception, and so forth, except the undoubted failure of modern science in its claims to universal rationality? How can one fail to see that the esotericism, elitism, inaccessibility, dehumanisation of – no doubt rigorous – sciences are precisely the counterpart of the false knowledge, easy illusions, passivity, maintained by these other, not so occult, sciences? Astrology today is but the other side of astrophysics. They coexist very well, cemented as they are by the dominant ideology, and the former will disappear only when the latter changes radically.

LIMITS AND PERSPECTIVES OF IDEOLOGICAL CRITICISM

I should like to conclude with two series of comments which concern the limits of the critical account I have given above.

Ideology within Scientific Activity

In the first place, it will be noticed that I have remained practically silent on the relations between ideology and the very act of scientific creation, that is the elementary and primordial activity of the scientific research worker(s), be it the work of the theoretician in front of the blank page or that of the experimentalist in front of the apparatus. What is — in terms of scientific knowledge in physics — the ideological component which results from this? I have already explained that such work is only a very limited part of scientific activity, even if it is the core of it. The ideological effects exercised by (and on) the political, economic, social and epistemological framework, in the centre of which scientific work (*strictu senso*) is immersed, seem to impose on it such a directive force that only little room for manoeuvre remains. Directions and priorities of scientific activities are decided at an ever more elevated level of the social apparatus and always further and further from the places of scientific work, especially in the case of sciences with economic and political implications as huge as today's physics. The organisation of scientific activity is determined in conformity with the general hierarchic model. Thus a criticism centred on the apparent forms of scientific discoveries does not seem essential, and in any case it would only have a significance on the basis of the more general criticisms developed in this chapter. To discuss the ideological consequences of the theory of fundamental particles, to understand whether hierarchic compound models are more reactionary than democratic bootstrap models, to evaluate the relations between Eastern philosophies and the eightfold way or bootstrap theory, to express a political judgement on the relative progressive qualities of the axiomatic field theory or of phenomenology based on the analytical properties of the S-matrix, though some have tried to do so, in my opinion means running the risk of becoming a prisoner of the very ideology one tries to denounce; such a procedure, deprived of any possibility of standing

outside the practice it wants to judge, cannot have any hold over practice, because it can only be heard by the physicists concerned — that is,
just those who can ignore it.

An internal and pertinent ideological criticism is certainly called for,
but on the essential condition that the criticism is articulated with the
mediations linking its particular sphere with other instances of social
life. Moreover, as I have just said, the target which presents itself to
purely internal criticism is extremely small. For example, I think that
there are not many ways, at present, of doing high-energy physics. The
Soviet example shows well that, once the principle of such an activity is
admitted — notwithstanding a bow in the direction of so-called 'Marxist
orthodoxy' — then, whether in the Soviet Union or in the United States,
it is a question of the same physics. The scarce information we possess
on the work being done in China — apparently little developed in this
field — suggests that, apart from some formal discussions on the mate-
rialistic merits of compound models, it is not easy to radically change a
point of view on the internal content of any science. One could say the
same of all the most fundamental research work, for instance in astrophysics, in general relativity and (therefore) in cosmology, though their
philosophical implications are considerable. However, it could happen
that clearer differences appear in sections of physics more closely
linked with economic production, such as low-energy nuclear physics;
research on controlled thermonuclear fusion (in this field the Soviets
seem to show more originality than the Americans); in solid-state
physics (where China could give us a surprise some day). But, once
more, it appears to me that the risk of dogmatism and schematism in
seeking to identify ideological influences on the forms of scientific
discovery is considerable.

In addition, bourgeois ideology itself has successfully opposed the
'neutrality' of scientific results and the lack of correlation with their
author's ideological and philosophical position to certain, though well-
intentioned, criticisms. For instance, the theory of relativity does not
seem to reflect Einstein's democratic humanism, quantum formalism
bears no mark of the aristocratic detachment of Dirac, quantum
principles of symmetry are unaffected by the political ideas and reactionary attitudes of Wigner or Gell-Mann, and so on and so forth.
Constructive attempts have been made to build heterodox physical
theories on an explicity ideological (sometimes allegedly 'Marxist')
basis; such was the case of the determinist sub-quantum constructions
of Bohm or Vigier. These attempts have so far failed, as Lysenko's

theory failed in biology. The reasons for this are clear if, as I have tried to demonstrate, ideological determinations operate institutionally much more than individually; that is to say, far up-stream or down-stream from the place where the flux of scientific production emerges as a delimited, formulated discovery.

For all that, I should not like to make anybody think that total indifference is recommended. To use the same metaphor, it might be extremely fruitful to study the underground movements of the flux before its resurgence, via the twisted paths of the scientist's unconscious. A psychoanalytical point of view, on condition that it remained critical and aimed at supplementing instead of supplanting the other aspects of ideological analysis, could reveal itself as very rich. It would throw a new light on the link between individual and subjective motivations of research workers and their objective collective role. Moreover, the growing and mutual influence of psychoanalytical and linguistic research could be relevant here. At the level we are considering, such approaches would perhaps permit the understanding of at least one bond, whose existence, if not its nature, seems clear. This bond, between ideology in general and scientific discovery in physical sciences in particular, materialises in the relation between both mathematical (formalised) and ordinary (heuristic) languages. This duality, which to me is fundamental to physics, poses a serious problem. It is much more than a simple question of terminology and thus of an apparently arbitrary choice between such-and-such a word of a language used to designate a physical concept, whose proper definition is attached to an abstract formalism.

It is not a question of a simple relation between *signifying* (the word) and *signified* (the concept), since, through its very pre-existence within current language (except in the case of absolute neologisms), the term chosen will suggest a series of well-determined (ideological in particular) connotations, which will react on the meaning then attributed to the concept and consequently on the manner it will be used and sometimes exploited. It would be worthwhile examining, to that effect, the vocabulary used in classical mechanics. Work, force, energy, power, and so on, are all words whose links with economic production are perfectly clear, and the ideological aspect of the choice in terminology can hardly be denied. Reciprocally, the exploitation of the names borne by physical concepts, through abuses of language, should be examined closely, and their propriety scrutinised; for instance, Marx's productive *forces*, labour *power*; or Freud's libidinal *energy*, to quote but a few major examples. A study of the terminology of contemporary physics would be the more interesting as this terminology reveals itself — in the light of epistemological criticism — as absolutely inadequate and clumsy, and very often tends to hinder the physics seriously. The theory of *relativity*, *uncertainty* relations, *observable* quantities: here is some froth on the surface of scientific writing, the result of previous obstacles met by the current and at the same time a cause for further disturbances. However, this kind of analysis is extremely difficult; it demands both very fine apparatus and — as said before — a very general framework. The question is wide open.

Ideological Crisis and Criticism

The very availability of an essay as this reflects the existence of a deep ideological crisis in the scientific milieu. This crisis is particularly obvious in the field of physics.[24] It is expressed, on the one hand, by a lack of motivation on the part of many young research workers, and, on the other hand, by the efforts of readjustment and self-justification on the part of the establishment. It is characterised by a serious loss of credibility in traditional values, which before had made it possible for research workers to create acceptable self-images. The esotericism of peak physics makes it more and more difficult to argue that knowledge in itself is in the 'interest of the whole of humanity'. The military and technical applications of more traditional branches undermine any prospect of using scientific progress for the benefit of humankind. It would be wrong, however, to think that such political or ideological views have led to the present situation. Far from being an explicit criticism that has started the crisis, it is the development of the crisis that has made criticism possible. Indeed, in the past, there were many scientists, and (great) physicists in particular, who were liberals, democrats or even 'left wing' (Einstein is an example); they were never driven to criticise the content of scientific activity. On the contrary, they lived with the idea that 'scientific method' supplied a means of universal analysis of reality, and that physicists, *because* they were physicists, were able, for instance, to understand social mechanisms.

Today, however, the extent of the division of labour in scientific work has considerably altered the image that research workers have of themselves. I shall deal with the case of the great majority of these workers, leaving out the simpler cases of scientists at the top of the hierarchy. Under its three essential forms of accentuated social hierarchy, specialisation and esotericism, this fragmentation of scientific tasks demands from researchers a subjective adjustment to their social function, which occurs at present in three different ways, reflecting three types of professional ideology.

(a) *Elitism.* This is the method adopted by 'brilliant young scientists' who arrived at research through prestigious channels such as the Higher Schools (Polytechnique or Normale Supérieure, in France) or famous universities of Britain or the United States. They generally hold a position in full-time research and may ignore the serious ideological and economic problems of teaching and relating to students. They are convinced of their own personal value and of the interest, in itself, of the notions they are trying to unveil. They play the games of intense competitiveness and productivity, which prevail, for instance, in particle physics. Their ascent in the hierarchy seems to them scientifically justified (and, as I have said before, so it is, ideologically).

(b) *Professionalism.* This is the present ideology of the largest number of research workers.[25] They think of themselves as having a profession 'like any other profession', 'rather well paid for not too much work'. They do not believe in their 'mission' but find that what they do is 'rather more interesting than something else'. And as long as they are paid for doing research work, they do not ask too many questions. The speed in which 'fashion' succeeds 'fashion' in research topics leaves them indifferent; they do not try to follow the rhythm and are quite contented to write just enough to fill in the yearly report of activity. The professional types, however, are more and more coming into sharp conflict though they minimise the amount of energy put into it — with the demands of 'profitability' and 'mobility' imposed on research workers by the bodies to which they belong, such as CNRS in France.

(c) *Criticism.* Still a small, yet increasing, number of scientists are less and less prepared to tolerate their alienation within a scientific system which is more and more obviously integrated into the social mechanisms of production, exploitation of work, political domination and oppression (I shall come back soon to the forms, limits and perspectives of this critique).

If this trend is developing today, it is less because of a sudden influence of external, ideological and political criticisms within scientific practice, than of the consequences of the increase of sharp inner contradictions between the traditional professional ideology and the subjective reality of scientific work. It was possible in the past to separate global, political and ideological positions from scientific practice. This practice was considered neutral by the conservatives, progressive by the left. If the latter did not ask deeper questions as to the nature of their work, it was because, although failing to control the motivations for support or the repercussions of their work, they could at least control the process itself. The major consequence of the mechanism of the division of labour, was, it seems to me, to end this control. Now, the production of knowledge, as the production of material goods, is fragmented. Average scientists do not even control the meaning of their own work. Very often, they are obscure labourers in theoretical computation or experimentation; they only have a very narrow perspective of the global process to which their work is related. Confined to a limited subject, in a specialised field, their competence is extremely restricted. It is only necessary to listen to the complaints of the previous generations' scientists on the disappearance of 'general culture' in science.

In fact, the case of physics is eloquent on the subject. One can say that, until the beginning of this century, the knowledge of an average physicist had progressed in a cumulative way, including progressively the whole of previous discovery. The training of physicists demanded an almost universal knowledge in the various spheres of physics. The arrival of 'modern' physics has brought about not only the parcelling of fields of knowledge, but also the abandonment of whole areas. I have already said that important sections of nineteenth-century physics are today excluded from the scientific knowledge of many physicists. Therefore the fields of competence are not only getting narrower, but some of them are practically vanishing altogether. If physicists no longer know about physics, *a fortiori* they know nothing about science! The idea of a 'scientific culture', of a 'scientific method', of a 'scientific spirit', which were common to all scientists and used to give them a large capacity for the rational understanding of all reality, have turned into huge practical jokes. True, some scientists have access to a global vision of their field or even of the social organisation of science and social ties, but that tends to depend solely on the position of power they occupy. The others, massively, are dispossessed of all mastery over their activity. They have no control, no understanding of its direction.

One result of the loss of control is the growing anarchy in scientific publications. There is a flood of papers, often of mediocre quality and on topics of no interest, except that of maintaining their authors in their positions. 90 per cent of scientific papers are never mentioned in further publications. The increase in the number of scientific journals makes for a good commercial business for the publishing houses, yet with all this massive production it is impossible to assemble a serious, large up-to-date documentation. Then, intensive and short-lived fashions flourish on such or such an idea launched by a big name. But the bone is abandoned half-eaten, abandoned when it gets too hard, and then the next one is tried. The path of theoretical physics has been lined for the last few years with incomplete, derelict buildings: peratisation, Regge's poles, bootstrap, unitary symmetries, current algebra, and so on, as many fashions as notions, certainly useful but with badly delimited concepts whose validity was partial and uncontrolled. The style of scientific work is becoming superficial, the direction is lost inside the global project; the research scientist works on a short-term basis and no longer knows the general significance of his or her production.

Here is a question of a real intellectual proletarianisation; scientists are as dispossessed of the products of their minds as workers of the products of their hands. The ideological crisis in science arises from the contradiction between this reality and the image, mentioned above, of a science considered as a general and global control of reality. Evidently, and of first importance, there are economic aspects of the crisis: insecurity of employment, demands for increased productivity, and so on. This description is particularly relevant in physics.

Yet the critical ideology, which today spreads into the scientific milieu, is far from having acquired a satisfactory strength or shape, particularly because it is torn apart by two opposite tendencies. One pushes towards externalisation, so as to place the weight of the criticism on extra-scientific institutions, on causes and consequences of scientific production; denouncing, for instance, the disastrous influence of capitalist monopolies or the role of the army. The second, by contrast tries to reduce the field of criticism to the visible core of scientific activity, to the explicit form of its results (I wrote earlier about the difficulty of apprehending it correctly). In both cases we witness the same phenomenon, that is the eclipse of the true judgement of scientific production.

One only has to see how these two tendencies coexist very well in the positions of the orthodox Western European communist parties; they will possibly discuss the validity of the concepts of contemporary physics — separately they will condemn the military use in Vietnam of the applications of that physics, but they will be opposed to every active criticism of such or such an American physicist, a genius perhaps, but also a Pentagon war collaborator. As to the most radicalised sections of the scientific milieu, they seem at the moment to have great difficulties in collectivising and socialising their criticisms, in particular for fear — justified no doubt — of falling victims again to more subtle forms of the dominant ideology, into expertise of counter-expertise, hierarchy of counter-hierarchy and establishment of anti-establishment.

In other words, if it is a question of proceeding to an internal criticism of scientific practice, a certain exteriority is absolutely necessary to prevent this criticism from being prisoner of the very trap it intends to warn against. For this, it seems to me necessary to arrive at a materialist conception of scientific activity which takes into account the whole social reality. The feeling of impotence and passivity engendered by technical achievements, and the vanishing of theoretical science from daily life, are not alien to scientists. In everyday life, average scientists hardly know how to deal with their technical environment (how to repair a car, and so on). Thus they find, at an individual level, the alienation and loss of control that I have described earlier as characteristic of the professional situation. Conversely, it is probably from being conscious of this dual incompetence that new forms of criticism could emerge. Thus, it should become possible to pose the problem of a 'science for the people' on a different level than that of rejection or immediate assimilation of 'peak' knowledge and technique, both equally impossible. Of course it is clear that the long-term abolition of the distinction between the scientific knowledge of an elite and the empirical knowledge of the mass, maintained in existence today by the dominant ideology, will demand a radical modification of science. The collective appropriation of scientific rationality will result in the disappearance of scientists as specialists. At this very moment, without giving up the indispensable struggle for political and ideological rights within the scientific institutions, but in order to wage the struggle with the necessary exteriority, radicalised physicists should spend some time out of their laboratories. Starting from science such as it is, hidden or denied in everyday life, a mass work would be possible, in which scientists would have to lose all elitism and paternalism, in order to exchange their theoretical knowledge for the practical knowledge of others, so that they re-acquire, together, the whole knowledge fossilised in objects and techniques whose apparent neutrality hide (but less and less well) their political and ideological role.

Notes and References

CHAPTER ONE

1. See, for example, M. Brown (ed.), *The Social Responsibility of the Scientist* (New York: Free Press, 1971).

2. S. Rose and H. Rose, in *The Social Impact of Modern Biology*, ed W. Fuller (London: Routledge & Kegan Paul, 1971).

3. See, for example, T. Roszak, *The Making of A Counter-Culture* (London: Faber & Faber, 1970); and such journals as *Environment* (United States) and *Your Environment.*

4. *Labó Contestation* (Lyons, 1970).

5. D. J. de Solla Price, *Little Science, Big Science* (New York: Columbia University Press, 1963).

6. H. Rose and S. Rose, *Science and Society* (London: Allen Lane, 1969).

7. N. Bukharin, *et al.*, *Science at the Crossroads* (London: Cass, 1971).

8. F. Engels, *Dialectics of Nature* (London: Lawrence & Wishart, 1940), V. I. Lenin, *Materialism and Empiriocriticism* (Moscow: Progress Publishers, 1964).

9. J. B. S. Haldane, *The Marxist Philosophy and the Natural Sciences* (London: Allen & Unwin, 1937).

10. J. Needham, *Science and Civilisation in China*, 4 vols (Cambridge University Press, 1954 – 71).

11. J. D. Bernal, *The Social Function of Science* (London: Routledge & Kegan Paul, 1939); *Science in History* (London: Watts, 1954).

12. See, for example, such diverse figures as E. B. Chain, *New Scientist* (October 1970); E. H. S. Burhop, 'Lecture to Technical University, West Berlin' (19 November 1971).

13. L. R. Graham, *Science and Philosophy in the Soviet Union* (New York: Knopf, 1972).

14. Z. Medvedev, *The Rise and Fall of T. D. Lysenko* (Columbia University Press, 1969).

15. H. J. Müller, *Out of the Night* (New York: Vanguard, 1935).

16. *Proceedings of the Engels Society*, nos 1 and 2 (1949), and no 7 (1952).

17. R. Williams, *The Long Revolution* (London: Chatto & Windus 1961).

18. P. Anderson, 'Components of the National Culture', *New Left Review* 50 (1968).

19. See, for example, *Theoretical Practice*, 3 – 4 (1971).

20. J. Mayer, in *Chemical and Biological Warfare*, ed. S. Rose (London: Harrap, 1968).

21. L. Fieser, *The Scientific Method* (New York: Reinhold, 1964).

22. J. Allen (ed.), *March 4th Scientists, Students and Society* (M I T Press, 1970).

23. Brown, *The Social Responsibility of the Scientist.*

24. See the various SESPA and related publications, the magazines *Science for the People, Science for Vietnam Newsletter,* and so on.

25. *Labó Contestation.*

26. Perche Si occupa il laboratorio Internazionale di Genetica e Biofisica. Collective of occupants (Naples, 11 May 1969).

27. Rose (ed.), *Chemical and Biological Warfare.*

28. *BSSRS Newsheet,* no. 1 (April 1969).

29. H. Rose and R. Stetler, *New Society* (25 September 1969).

30. S. Rose, *The Himsworth Memorandum* (London: BSSRS, 1970).

31. See the newsheet of the Edinburgh Society for Social Responsibility in Science.

32. See the newsheet of the Cambridge Society for Social Responsibility in Science, *Science or Society.*

33. *Cambridge Society for Social Responsibility in Science Bulletin;* reprinted in *The Biological Basis of Behaviour,* ed. N. Chalmers, R. Crawley and S. Rose (London: Harper & Row, 1971); see also the later book edited by K. Richardson and D. Spears, *Race, Culture and Intelligence* (Harmondsworth: Penguin, 1972); and the pamphlet *Racism, IQ and the Class Society* issued by the Campaign on Racism, IQ and the Class Society in 1974. See also *The Political Economy of Science,* chapter 7.

34. H. J. Eysenck, *Race, Intelligence and Education* (London: Temple Smith, 1971).

35. W. Fuller (ed.) *The Social Impact of Modern Biology* (London: Routledge & Kegan Paul, 1971).

36. G. Wick, *New Scientist;* also Solidarity pamphlet, *Soldier Technicians or Irresponsible Scientists?* (1970).

37. *BSSRS Newsheet* (April – May 1972).

CHAPTER TWO

1. D. Joravsky, *The Lysenko Affair* (Harvard University Press, 1970).

2. Z. Medvedev, *The Rise and Fall of T. D. Lysenko* (New York: Columbia, 1969).

3. L. K. Prezent, an attorney by training, joined Lysenko first as a polemicist and then did some experimental work. He was an especially strident, dogmatic and abusive participant in the debates.

4. J. B. Lamarck (1744 – 1829), author of *Zoological Philosophy*, argued that evolution occurs through the inheritance of acquired adaptive responses to the environment.

5. B. Hessen, Director of the Moscow Institute of Physics, made an ambitious, but only partly successful, attempt to give a materialist interpretation of the early history of Western physical science in an article entitled 'The Social and Economic Roots of Newton's "Principia" ', *Science at the Crossroads* (London: Kniga, 1931). J. D. Bernal, British Marxist physicist, gave a Marxist interpretation of the history of sociology science in his two main historical works, *The Social Function of Science* (London: Routledge & Kegan Paul, 1939) and *Science in History* (London: Watts, 1954).

6. Some examples of Lysenkoist studies, showing the range of work, are: V. F. Khitrinsky, 'On the Possibility of Directing the Segregation of the Hybrid Progeny of Wheat'; G. I. Lashuk, 'Changes in the Dominance of Alkaloid Characters in Interspecific Hybrids of *Nicotiana*; Sisakian's work on the transmission of enzymatic activity by grafting; Turbin's study in which a multiple recessive tomato was pollinated with a mixture of pollen each carrying a single dominant and gave some offspring with two dominant phenotypes; Avakian's use of foreign pollen to overcome self-sterility in rye; Olshanksy's work on the effect of conditions in the F_1 generation on the segregation ratio in the F_2; Isayev's claim that the offspring of graft hybrids sometimes show the same kind of segregation met with in ordinary sexual crosses; and Glushchenko's book on vegetative hybridisation. A general review may be found in P. S. Hudson and R. H. Richens, *The New Genetics in the Soviet Union* (Cambridge: Imperial Bureau of Plant Breeding and Genetics, 1946).

7. M. G. Kains, *Plant Propagation – Greenhouse and Nursery Practice* (New York: Orange Judd, 1916) p. 175.

8. I. I. Schmalhausen, *Factors of Evolution* (Philadelphia: Blakiston, 1949).

9. T. Dobzhansky, *Genetics and the Origin of Species*, 3rd edn (New York: Columbia, 1951).

10. K. H. W. Klages, *Ecological Crop Geography* (New York: Macmillan, 1949).

11. In regions of poor summer rainfall, seed planted in the spring may not achieve sufficient growth before the dry season. Thus, some crops, notably wheat, have a 'winter' variety in which seeds are planted in the autumn, undergo a small amount of growth, then over-winter as very young seedlings, and then start to grow again immediately in the spring, thus getting a longer total growing season. Such 'winter' varieties, however, are subject to catastrophic loss if the winter is unduly severe. 'Vernalisation' is a process of chilling and wetting the seed of 'winter' varieties, but then planting these 'vernalised' seeds in the spring. The seed will successfully complete their growth cycle without the usual winter chilling, and also avoid exposure to the hazard of frost. The question remains open whether the advance in growth over normal spring varieties in fact results in increased yield. While the process was known in the nineteenth century, Lysenko adopted it and expanded its use to a whole variety of crops and situations.

12. Also called 'super-early sowing', this involved planting seed in unploughed fields just after the winter snow had melted.

13. Quoted by Joravsky, *The Lysenko Affair*, p. 84.

14. A. Weismann (1834–1914) emphasised the distinction between germplasm and somatoplasm, the continuity of the germplasm from generation to generation, and its immutability.

15. For these and other similar experiments and theoretical arguments relevant to Lamarckist interpretations, the following bibliography is representative: N. J. Berrill and C. K. Liu, 'Germplasm, Weismann and Hydrozoa', *Quarterly Review of Biology*, 23 (1948) pp. 124–32; H. L. Bolley, 'Indication of the Transmission of an Acquired Character in Flax', *Science*, 66 (1927) pp. 301–2; J. T. Cunningham, 'Evolution of the Hive-bee', *Nature*, 125 (1930) p. 857; L. Daniel, 'The Inheritance of Acquired Characters in Grafted Plants', *Proceedings of the International Congress of Plant Sciences*, 2 (1926) pp. 1024–44; W. H. Eyster, 'The Effect of Environment on Variegation Patterns in Maize Endocarp', *Genetics*, 11 (1926) pp. 372–86; H. Federley, 'Weshalb lehnt die Genetik die Annahme einer Vererbung erworbener Eigenschaften ab?', *Paleontologische Zeitschrift*, 11 (1929) pp. 287–317; J. E. Finesinger, 'Effect of Certain Chemical and Physical Agents on Fecundity and Length of Life, and on their Inheritance in a Rotifer', *Lecane*

inermis (Bryce)', *Journal of Experimental Zoology*, 44 (1926) pp. 63 – 94; M. F. Guyer, 'The Germinal Background of Somatic Modifications', *Science*, 71 (1930) pp. 109 – 76; J. W. H. Harrison, 'Experiments on the Egg Laying Instincts of the Sawfly, *Pontamia salicis* Christ., and their Bearing on the Inheritance of Acquired Characteristics with some Remarks on a New Principle of Evolution', *Proc. Roy. Soc. Lond. B.*, 101 (1927) pp. 115 – 26; G. Klebs, 'Alterations in the Development and Forms of Plants as a Result of Environment', *Proc. Roy. Soc. Lond. B.*, 84 (1910) pp. 547 – 58; S. Konsuloff, 'Über die Dauermodifikationen den tierischen Gewebe', *Zeitschrift der Geselschaft Exp. Med.*, 89 (1933) pp. 177 – 82; P. Lesage, 'Sur la précocité: étapes du caractère provoqué, au caractère hérité définitivement fixé. Application à la prediction de primeurs', *Comptes Rendues de l'Académie d'Agriculture*, 182 (5) (1924); P. Lesage, 'Sur la précocité provoquée et heritée dans le *Lepidium sativum* après la vie sous chassis', *Revue générale de botanique*, 38 (1926) pp. 65 – 86; E. W. MacBride, 'Habit: The Driving Force in Evolution', *Nature*, 127 (1931) pp. 933 – 44; F. Nopsca, 'Heredity and evolution', *Proceedings of the Zoological Society of London*, 2 (1926) pp. 633 – 65 – 'the foregoing lines all lead to the idea of heredity being regulated by two antagonistic factors. The conservative factor evidently is the constancy of the germplasma. . . the other modernising factor that enables all evolution to go on must consist of the cooperation of all these different physiological factors by which the germplasma is either directly affected or indirectly altered by the medium of some hormone.' W. Pfeffer, *The Physiology of Plants*, 2nd edn (Oxford: Clarendon Press, 1900) – 'External conditions act not so much as direct formative, as indirect inducing agents and thus produce vital changes leading to an attainment of new hereditary peculiarities.' p. 83; J. M. Reynolds, 'On the Inheritance of Food Effects in the Flour Beetle, *Tribolium destructor*', *Proceedings of the Royal Society of London* (B) 132 (1945) pp. 438 – 51; D. E. Sladden and H. R. Hewer, 'Transference of Induced Food Habit from Parent to Offspring III', *Proceedings of the Royal Society of Edinburgh* B, 126 30 – 44; F. J. Stevenson, 'Potato Breeding Genetics and Cytology: Review of Literature of Interest to Potato Breeders', *American Potato Journal*, 25 (1948) pp. 1 – 12; A. H. Sturtevant, 'Can Specific Mutations be Induced by Serological Methods?', *Proceedings of the National Academy of Sciences of the United States*, 30 (1944) pp. 176 – 8; P. M. Suster, 'Erblichkeit aufgezwungener Futterannahme bei *Drosophila repleta*', *Zoologischer Anzeiger*, 102 (1933) pp. 222 – 4; T. Swarbrick, 'Root

Stock and Scion Relationship. Some Effects of Scion Variety upon the Root Stock', *Journal of Pomology and Horticultural Science*, 81 (1930) pp. 210 – 28; H. M. Vernon, 'The Relations between the Hybrid and Parent Forms of Echinoid Larvae', *Philosophical Transactions of the Royal Society of London* B, 190 (1898) pp. 465 – 529; K. S. Wilson and C. L. Withner Jr, 'Stock – Scion Relationships in Tomatoes', *American Journal of Botany*, 33 (1946) pp. 796 – 801.

16. Berrill and Liu, 'Germplasm, Weismann and Hydrozoa'.

17. F. Griffith, 'The Significance of Pneumococcal Types', *Journal of Hygiene*, 27 (1928) p. 113.

18. B. Zavadovsky, 'The "Physical" and the "Biological" in the Process of Organic Evolution', in *Science at the Crossroads* (London: Kniga, 1931).

19. See Joravksy, *The Lysenko Affair*, p. 427, for relevant Soviet literature.

20. E. H. Carr, *The Bolshevik Revolution*, vol. 2 (London: Macmillan, 1952).

21. R. Levins, *Evolution in Changing Environments* (Princeton University Press, 1968).

22. R. C. Lewontin, *The Genetic Basis of Evolutionary Change* (New York: Columbia University Press, 1974).

CHAPTER THREE

1. C. Zetkin, 'Women, Marriage and Sex', *Reminiscences of Lenin* (New York: International Publishers, 1934) p. 53.

2. See R. Dunbar, 'La caste et la classe', *Partisan*, no. 54 – 5 (1973) p. 50; C. Dupont, 'L'Ennemi principal', *Partisan*, no. 54 – 5 (1973) p. 157; and an Italian collective, *Etre exploitées*, (Paris: des Femmes, 1974).

3. Figures cited by J. Verdes-Leroux, *les Temps Modernes* (August – September 1974) p. 270.

4. Ibid.

5. Ibid.

6. Zetkin, *Reminiscences of Lenin*.

CHAPTER FOUR

1. R. Barthes, *Le Plaisir du texte* (Paris: de Seuil, 1973).

2. B. Jurdant, 'Les problèmes théoriques de la vulgarisation scientifique', Thèse de Doctorat (Strasbourg: Université Louis Pasteur, 1973).

3. Sarah Fienas, 'Journée d'une chercheuse dans un labo', *Liberation* (18 June 1974).

4. L. M. Bachtold and E. E. Werner, 'Personality Characteristics of Women Scientists', *Psychological Reports*, 31 (1972) pp. 391 – 6 and references therein.

5. E. Maccoby, 'Woman's Intellect', in *The Potential of Woman*, ed. S. M. Farber and R. L. Wilson (New York: McGraw-Hill, 1963). *Impact of Science on Society*, XX (1) (1970). 'Sex Differences in Intellectual Functioning', in *The Development of Sex Differences*, ed. E. Maccoby (Stanford University Press, 1956).

6. F. Clemente, 'Early Career Determinants of Research Productivity', *American Journal of Sociology*, 79, 409 (1974).

7. B. Bettelheim in *Women and the Scientific Professions. The MIT Symposium on American Women in Science*, ed. J. A. Mattfeld and C. G. van Aken (MIT Press, 1965).

8. A. Moles, *Entretiens en marge de la science nouvelle* (The Hague: Mouton, 1963).

9. J. K. Galbraith, *Le nouvel état industriel* (Paris: Gallimard, 1968).

10. R. Richta, *La civilisation au carrefour* (Paris: Anthropos, 1972).

11. H. Marcuse, *One Dimensional Man* (London: Routledge & Kegan Paul, 1964).

12. J. Baudrillard, *Le miroir de la production ou l'illusion critique du matérialisme historique* (Paris: Casterman, 1973).

13. F. Engels, 'The Origin of the Family', *Marx and Engels Selected Works* (London: Lawrence & Wishart, 1968).

14. International Labour Office, 'Shortage of Highly Qualified Engineers and Scientists', *International Labour Review* (1957) pp. 588 – 603.

15. V. Terechkova, *Impact of Science on Society*, XX (1) (1970).

16. C. Holden, 'NASA Satellite Project: the Boss is a Woman', *Science*, 179, 48 (1973).

17. S. Rajender, 'Women in Academia: the Plight and its Perpetuation', *Chemical Technology* (August 1973) p. 475.

18. J. Bernard, *Academic Women* (Pennsylvania State University Press, 1964).

19. See *Women and the Scientific Professions*, ed. Mattfeld and van Aken.

20. H. Astin, *The Woman Doctorate in America: Origins, Careers and Family* (New York: Russell Sage Foundation, 1969); A. H. Cook,

'Sex Discrimination at Universities: an Ombudsman's View', *American Association of University Professors Bulletin*, 58 (1972) pp. 279-82; G. Ezorsky, 'The Fight over University Women', *New York Review* (16 May 1974); 'Medical Education: those Sexist Putdowns may be Illegal', *Science*, 184, 449 (1974) and issue reviewed therein: 'Why would a Girl go into Medicine? Medical Education in the U.S., a Guide for Women'; M. S. White, 'Psychological and Social Barriers to Women in Science', *Science*, 170, 413 (1970); N. F. Goldsmith, 'Women in Science: Symposium and Job Mart', *Science*, 168, 1124 (1970); P. A. Graham, 'Women in Academia', *Science*, 169, 1284 (1970); M. A. Ferber and J. W. Loeb, 'Performance, Rewards and Perception of Sex Discrimination among Male and Female Faculty', *American Journal of Sociology*, 78 (1973) pp. 995-1001; C. Holden, 'NASA: Sacking of Top Black Women Stirs Concern for Equal Employment', *Science*, 182, 805 (1973); 'Les jeunes femmes diplômées d'université, leur mariage, leur vie professionnelle, leurs problemes', *Review de l'Institut de Sociologie Solvay*, 33 (1) (1973) pp. 103-56.

21. See Ezorsky, 'The Fight over University Women'; and Holden, 'NASA: Sacking of Top Black Women'.

22. Astin, *Woman Directorate in America*.

23. N. T. Dodge, *Women in the Soviet Economy: their Role in Economic, Scientific and Technical Development* (Baltimore: Johns Hopkins Press, 1966).

24. Hongqi (December 1973) cited by C. Broyelle in *Chine 74* (Association des Amitiés franco-chinoises).

25. S. de Beauvoir, *Le Deuxième Sexe* (Paris: Gallimard, 1949).

26. *La femme dans la société: son image dans les differents milieux sociaux* (Éditions du CNRS, 1967).

27. A. M. Rocheblave Spenle, *Les rôles masculins et féminins* (Paris: Éditions Universitaires 1964).

28. D. Lagache, Ibid. 'Preface'.

29. S. Freud, 'Femininity', in *New Introductory Lectures in Psychoanalysis* (New York: Hogarth Press, 1964).

30. S. de Beauvoir, *Le Deuxième Sexe*.

31. K. Millett, *Sexual Politics* (London: Abacus, 1971).

32. S. Freud, 'Civilized Sexual Morality and Modern Nervousness', *Collected Papers* (1908); cited in Rocheblave-Spenle, *Les rôles masculins et féminins*.

33. S. Moscovici, *La psychoanalyse, son image et son public* (Paris: PUF 1961).

34. J. Lacan, *Télévision* (Paris: de Seuil, 1974).

35. Engels, *Origin of the Family*.

36. K. Marx, *1844 Manuscripts* (Paris: Gallimard, 1963).

37. J. Ortega y Gasset, *The Revolt of the Masses* (New York: Norton, 1932).

38. J. R. Cole and S. Cole, 'The Ortega Hypothesis', *Science*, 178, 368 (1972).

39. J. Monod. *Le Hasard et la Necessité* (Paris: de Seuil, 1970).

40. Cynthia Fuchs Epstein. 'Success among Women', *Chemical Technology* 8 (1973).

41. J. Baudrillard, *Pour une critique de l'économie politique du signe* (Paris: Gallimard, 1972).

42. J. Lacan, *Écrits* (Paris: de Seuil, 1971).

43. *Women and the Scientific Professions*, ed. Mattfeld and van Aken.

44. Astin, *Woman Doctorate in America*.

CHAPTER FIVE

1. Earlier presentations of some rather similar thoughts have appeared in *University of Hongkong Gazette,* 21, no. 5 (1974) pt 1, 69; and in *Impact of Science on Society,* 25, no. 1 (1975) pp. 45, 49. 'Dilemmes de la Science et de la Médécine Modernes — le Remède est-il Chinois?'

2. I use this phrase to designate what I believe to be the immediate goal of human social evolution. It was a favourite one on the lips of Conrad Noel, parish priest of Thaxted in Essex, among the greatest of my teachers.

3. An account of those days is contained in *Science Outpost* (London: Pilot Press, 1948), and *Chinese Science* (London: Pilot Press, 1945).

4. J. Needham, *Science and Civilisation in China* (Cambridge University Press, 1954 onward).

5. This phrase echoes the 'big science' and 'little science' antithesis, and the 'background noise of low technology', discussed by Derek J. de S. Price in *Science since Babylon* (Yale University Press, 1961) and other works.

6. In *The Decline of the West* (London: Allen & Unwin, 1926), a book famous in its time, Oswald Spengler pictured the successive cultural entities, for example, Ancient Egyptian, Islamic, Indian,

Magian, Chinese, Faustian, and so on, as separate organisms having very little interaction between each other, and arising, flourishing, decaying and dying like incommensurable plant or animal forms. A revaluation has been done recently by N. Frye in *Daedalus* (Winter 1974), and *Proceedings of the American Academy of Arts and Sciences*, 103, no. 1 (1971).

7. The story has been told in detail by J. Needham, L. Wang D. J. de S. Price in *Heavenly Clockwork* (Cambridge University Press, 1960). On the astronomical background see *Science and Civilisation in China*, vol. 3, pp. 229 ff., 339 ff.

8. *Science and Civilisation in China*, vol. 3, pp. 409 ff.

9. V. F. Weisskopf, *Physics in the Twentieth Century: Selected Essays* (M I T Press, 1972) p. 94. Quite recently there was some correspondence in *The Times* about the pulsar mentioned here, where it was said that the Chinese had described the bright new star in 1954, so of course I was unable to resist the temptation of writing to point out that this misprint had knocked off nine hundred years from their achievement.

10. See *Science and Civilisation in China*, vol. 3, pp. 473 ff. More fully. Ho Ping-Yü & J. Needham, Ancient Chinese Observations of Solar Haloes and Parhelia', *Weather*, 14 (1959) p. 124.

11. See J. Needham, 'The Pre-natal History of the Steam-Engine', Newcomen Centenary Lecture, *Transactions of the Newcomen Society*, 35 (1963) p. 3.

12. All details will be found in *Science and Civilisation in China*, vol. 4, pt 2; and *Clerks and Craftsmen in China and the West* (Cambridge University Press, 1970), but new discoveries are still being made.

13. This last development is attested for this date by a passage in the *Loyang Chhieh-Lan Chi* (Description of the Buddhist Temples and Monasteries in Loyang), written by Yüang Hsüan-Chih in +547. It was first noticed by Dr William Jenner.

14. See Hartwell's important papers, for example 'The Revolution in the Chinese Iron and Coal Industries during the Northern Sung'. *Journal of Asian Studies*, 21 (1962) 153; 'A Cycle of Economic Change in Imperial China; Coal and Iron in the Northeast, +750 to +1350, *Journal of Economic and Social History of the Orient*, 10 (1967) 102; 'Markets, Technology and Enterprise in the Development of the Eleventh-Century Chinese Iron and Steel Industry'. *Journal of Economic History*, 26 (1966) 29.

15. See P. Huard & Huang Kuang-Ming (M. Wong), 'La Notion

de cercle et la Science Chinoise', *Archives Internationale d'Histoire des Sciences*. 9 (1956) 111.

16. The literature of acupuncture is inchoate, difficult of assembly and full of pitfalls, especially for those who have no linguistic access to Chinese, but F. Mann's *Acupuncture, the Ancient Chinese Art of Healing* (London: Heinemann, 1962; often reprinted) is a good beginning.

17. It has given us no small amusement to collect examples of these reactions. Obliged at last to acknowledge a non-European achievement, Western historians of science and technology retire in good order by redefining (to their own advantage) what the achievement really ought to be considered to be. The first of all clock escapements may be Chinese, but the only true, the only important, escapement, was the verge-and-foliot of course (*Science and Civilisation in China*, vol. 4, pt 2, p. 545). The magnetic needle may have been used first by Chinese mariners, but of course only the wind-rose attached to the magnet is the *véritable boussole* (Ibid. vol. 4, pt 3, p. 564). Chinese seamen first mounted axial rudders, no doubt, but of course there could be no true rudder without a stern-post (Ibid. vol. 4, pt 3, p. 651). This is what we call the Department of Face-Saving Redefinitions.

18. See *Science Outpost*, pp. 257–8.

19. *Science and Civilisation in China*, vol. 3, pp. 448 ff. Also J. Needham, 'Chinese Astronomy and the Jesuit Mission, an Encounter of Cultures', China Society Lecture (London, 1958). The change was from 'The West's New Calendrical Science' to 'The New Mathematics and Calendrical Science'.

20. See J. Needham, 'The Roles of Europe and China in the Evolution of Oecumenical Science', *Advancement of Science*, 24 (1967) 83; reprinted in *Clerks and Craftsmen*, p. 396.

21. *Science and Civilisation in China*, vol. 1, p. 19; *Time, The Refreshing River* (London: Allen & Unwin, 1943).

22. See A. Arber, *Herbals, their Origin and Evolution* (Cambridge University Press, 1912); K. F. W. Jessen, *Der Botanik in Gegenwart und Vorzeit* (Leipzig, 1864, reprinted Waltham, Mass., Chronica Botanica, 1948).

23. For brief introductions, see S. Cotgrove, 'Anti-Science', *New Scientist* (July 1973) p. 82; 'Objections to Science', *Nature*, 250 (1974) p. 764.

24. T. Roszak, *The Making of a Counter-Culture: Reflections on the Technocratic Society and its Youthful Opposition* (London: Faber & Faber, 1971).

25. T. Roszak, *Where the Wasteland Ends: Politics and Transcendence in Post-Industrial Society* (London: Faber & Faber, 1972 – 3).

26. Roszak, *Counter-Culture*, pp. 205 ff.

27. Ibid. pp. 217 ff.

28. Ibid. p. 222.

29. Ibid. pp. 227 ff. This is the opposite of what I quote below from the *Kuan Yin Tzu* book and the Gospels.

30. Roszak, *Wasteland*, p. 31.

31. See J. Ellul, *The Technological Society* (London: Cape, 1965).

32. See W. Leiss, *The Domination of Nature* (New York: Brazillier, 1972).

33. Roszak, *Wasteland*, p. 168.

34. For a recent study by a biologist of the ethics of animal experimentation, see C. Roberts, *The Scientific Conscience* (Fontwell, Sussex: Centaur Press, 1974). See also M. H. Pappworth, *Human Guinea-pigs: Experimentation on Man* (London: Routledge & Kegan Paul, 1967).

35. One has only to mention the Vietnam War in this connection, with its fragmentation bombs, napalm, computer analysis of terrain, vacuum asphyxia weapons, defoliants, and so on.

36. This whole subject has been the theme of C. H. Waddington's Bernal Lecture at the Royal Society, 'New Atlantis Revisited', *Proceedings of the Royal Society*, 190 (1975) pp. 301 – 14.

37. I call to mind especially here a book which had great influence on me when I was young, R. G. Collingwood, *Speculum Mentis* (Oxford: Clarendon Press, 1924). Current discussions are numerous – for example, T. Roszak, 'The Monster and the Titan; Science, Knowledge and Gnosis', *Daedalus* (Summer 1974) p. 17, together with the following article by S. Weinberg, 'Reflections of a Working Scientist' *Daedalus* (Summer 1974); also issued as *Proceedings of the American Academy of Arts and Sciences*, 103 (1974) no. 3.

38. *Wasteland*, pp. 106, 189.

39. Ibid. pp. 74 ff.

40. See, for example, K. Raine, *Blake and Tradition*, 2 vols (Princeton University Press, 1968); J. B. Beer, *Blake's Humanism* (Manchester University Press, 1968). In *Vala, or the Four Zoas*, G. E. Bentley has reconstructed the poem from Blake's chaotic notes, and given a facsimile reproduction of it (Oxford University Press, 1963).

41. Weisskopf, *Physics in the Twentieth Century*, p. 58.

42. Ibid. p. 349. Similar ideas have been expressed by T. R.

Blackburn, 'Sensuous-Intellectual Complementarity in Science', *Science*, 172 (1971) p. 1003.

43. Ibid. p. 351.

44. Ibid. p. 364.

45. *Revelations of Divine Love*, excerpts edited by L. Sherley-Price (London: Mowbray, 1962).

46. See F. Crick, *Of Molecules and Men* (University of Washington Press, 1968); B. F. Skinner, *Beyond Freedom and Dignity* (London: Cape, 1972); Jacques Monod, *Chance and Necessity* (London: Cape, 1972). See also the critical enquiry on this by eleven writers edited by J. Lewis, *Beyond Chance and Necessity* (London: Lawrence & Wishart, 1974). To the same tradition of 'scientific positivism' and reductionism belong Desmond Morris, *The Naked Ape* (London: Cape, 1968) and Peter Medawar's *The Art of the Soluble* (London: Methuen, 1967).

47. Affluence without altruism – the failure of the 'Welfare State'.

48. See D. W. Y. Kwok, *Scientism in Chinese Thought, 1900 to* (New Haven: Bill & Tennen, 1965).

49. The Chinese talk about the love and service of 'class brothers and sisters' but this is what they really mean.

50. It is of much interest that a Persian philosopher of science, Said Husain Nasr, has recently also inveighed against this, from the standpoint of Islam. A critique of his interesting books, *Science and Civilisation in Islam* (Harvard University Press, 1968), and *The Encounter of Man and Nature: the Spiritual Crisis of Modern Man* (London: Allen & Unwin, 1968) will be found in *Science and Civilisation in China*, vol. 5, pt 2, pp. xxiv ff. His complaint against science echoes that of Roszak, but he has less appreciation of the validity of science as a form of experience in its own right. One may not like some of its attitudes and blindnesses – *mais c'est la nature de l'animal*.

51. *Wasteland*, pp. 109 ff.

52. On all this, see E. Bevan, *Holy Images* (London, 1940).

53. This is a famous Muslim expression. When the Muslims in their great expansion encountered Hindus or Buddhists they were liable to insist on conversion to Islam at the point of the sword, but Jews and Christians were tolerated as minorities, even if oppressed, because they also accepted the Old Testament as a sacred book.

54. A. Waley, *The Way and the Power*, p. 179 (London: Allen & Unwin, 1934).

55. See the recent lecture by B. T. Feld, 'Doves of the World,

Unite!', *New Scientist* (December 1974) p. 910.

56. See A. Lovins, *Nuclear Power: Technical Bases for Ethical Concern*, Friends of the Earth, London.

57. See 'Genetics and the Quality of Life', *Study Encounter*, 10 no. 1 (1974); S.E. Report no. 53, World Council of Churches, Geneva.

58. This is called 'amniocentesis', the cytological examination of cells sampled by biopsy from the embryo and its membranes.

59. See the Ciba Symposium, *Law and Ethics of A.I.D. and Embryo Transfer* (London: Churchill, 1973).

60. On all these questions, see *Our Future Inheritance; Choice or Chance?*, ed. A. Jones & W. F. Bodmer, a study by a British Association Working Party (Oxford University Press, 1974).

61. See the Trueman Wood Lecture of C. H. Waddington, 'Genetic Engineering', *Journal of the Royal Society of Arts*, 123 (1975) p. 262. My old friend and collaborator takes a refreshingly cool view of the dangers before us, because of the immense expense which researches in such embryology and genetics involve, and the consequent certainty of public scrutiny. I am not so optimistic, for two reasons: (*a*) the possible activities of totalitarian states; and (*b*) the doubt that public scrutiny — or debate — necessarily leads to right ethical policy. As Waddington himself says, 'one wonders whether we are intellectually, emotionally, or morally, prepared to face such choices'.

62. This was the great discovery of the 'transforming principle' (free genes) by Avery and others. Its importance in relation to the solution of the problem of the genetic code has been clearly brought out by R. Olby in his remarkable book *The Path to the Double Helix* (London: Macmillan, 1974).

63. This was the Berg Conference at Asilomar, on which see *Nature*, 254 (1975) p. 6. Self-imposed controls for research involving genetic-transfer techniques had already been adumbrated in the Ashby Report (British Government), *Nature*, 253 (1975) p. 295.

64. The opening of such a laboratory by Imperial Chemical Industries at Runcorn was announced in *The Times* (6 June 1975).

65. One of the most obvious, and widely debated, is of course that of euthanasia. On all these questions, see G. Leach, *The Biocrats* (London: Cape, 1970).

66. See the Ciba Symposium, *Law and Ethics of Transplantation* (London: Churchill, 1966).

67. H. G. Wells's prophetic *Island of Dr Moreau* (London: Heinemann, 1896) has become only too real a possibility.

68. See R. G. Edwards, 'Fertilisation of Human Eggs *in Vitro*: Morals, Ethics and the Law', *Quarterly Review of Biology*, 49 (1974) p. 3. Also R. G. Edwards & D. J. Sharpe, 'Social Values and Research in Human Embryology', *Nature*, 231 (1971) p. 87.

69. My review in *Scrutiny* (1932) has been reprinted in *Aldous Huxley, the Critical Heritage*, ed. D. Watt (London, 1975) p. 202. It was closely in tune with what is now being said here more than forty years later.

70. See the Ciba Symposium, *Civilisation and Science: in Conflict or Collaboration?* (London: Churchill, 1972).

71. See, for example, his paper 'The Exhaustion of Possibilities of Theoretical Science in History and its Reasons', *Proceedings of the XIVth International Congress of the History of Science* (Tokyo, 1974); and numerous papers in the *Science Reports of the Society for Research in Theoretical Chemistry* (1955 onwards). Others have been thinking along similar lines, notably F. Capra, as in 'Modern Physics and Eastern Philosophy', *Human Dimensions*, 3, no. 2 (1974) p. 3.

72. John 1, 9.

73. Is it not a remarkable fact that the charismatic leader of the eight hundred million 'black-haired people', Mao Tse-tung, should be a social and ethical philosopher, not a military man? Plato's famous remark would be relevant here, though it does not seem easy to point to any European parallel. The contrary is only too obvious, from Alexander the Great to Napoleon, Hitler and Mussolini. And when philosophers or natural philosophers did appear at the top in Europe, as perhaps Alfonso X of Castile or Rudolph II in Prague, their reigns were not very successful. In the Chinese case we have a millennial background of respect for scholars and thinkers. I am indebted for this point to a conversation in Montreal with Ronald Melzack and Elizabeth Fox.

74. 'In everything you do, let it be done for people.'

75. At the present time, in China, Confucianism is unpopular and the good points of the Legalists are being rediscovered. This is quite justified because for far too long certain aspects of Confucius' teachings, especially the subjection of women, had been taken in a 'fundamentalist' way, without historical criticism of his acceptance of feudal or proto-feudal society. But like all the other Chinese sages he gave his ethics a naturalist, not a supernatural, basis; and he certainly propagated the doctrine that every man who could profit by education should have it, regardless of birth or wealth. Mo Ti, with his 'universal love', is now

widely appreciated, and the Taoists generated a wisdom implicit in most of what all Chinese instinctively do.

76. *Kuan Yin Tzu*, Ch. 7. p. 1a.

77. Luke 4, 3 – 13.

78. See *Science and Civilisation in China*, vol. 5, pt 2, p. 86.

79. See the interesting paper of W. A. Engelhardt, 'Hierarchies and Integration in Biological Systems', *Bulletin of the American Academy of Arts and Sciences*, 27 no. (4 (1971) 11).

80. Exactly the same point of view is put by V. F. Weisskopf in 'The Frontiers and Limits of Science', *Bulletin of the American Academy of Arts and Sciences*, 28, no. 6 (1975) 15.

81. For a wide-ranging study of labour and its ethics see the remarkable survey and bibliography by Gene Weltfish, *Work: an Anthropological View*, Module 9 – 065 issued by the Empire State College, University of New York (New York: Saratoga Springs, 1975).

82. China has had to mobilise a vast society and solve innumerable institutional problems, to create a viable educational system, to gain access to modern technology and to provide health services for everyone; but the success of all this depends on the acceptance of a universal value system based on solidarity, revolutionary mentality and the primacy of moral over material motives. See N. Ganière, 'The Process of Industrialisation of China, an Analytical Bibliography', O.E.C.D. Development Centre Working Document CD/TI (74) O; and C. G. Oldham, 'Scientific and Technological Policies', in *China's Developmental Experience*, ed. M. Oksenberg (New York: Academy of Political Science, 1973) pp. 80 ff.

83. See the interesting interview with Charles Bettelheim of the École Pratique des Hautes Etudes in Paris, 'Economics and Ideology', in *China Now*, no. 52 (1975) p. 9.

84. See L. E. Björk, 'An Experiment in Work Satisfaction', *Scientific American*, 232, no. 3 (1975) p. 17; also Gene Weltfish, *Work: An Anthropological View*.

85. Long ago Stuart Chase, in his remarkable book *Men and Machines* (London: Cape, 1929) enumerated the many different types of contact between them, ranging from the beneficial through the neutral to the dangerous and the lethal. See *Time, the Refreshing River*, p. 134. Here might be mentioned the extremely valuable 'Bibliography of the Philosophy of Technology' prepared by C. Mitcham & R. Macey, *Technology and Culture* 14, no. 2 (1973) pt 2.

86. See the important paper by Denis Goulet, 'Le Monde du Sous-

Developpement: une Crise de Valeurs', *Comptes Rendus de la Réunion de l'Association Canadienne des Études Asiatiques* (Montreal, 1975).

87. *Lectures on Conditioned Reflexes* (London, 1941) vol. 2, p. 53; quoted in *Time, the Refreshing River* (London: Allen & Unwin, 1943) pp. 156–7.

88. Of course all this is independent of the political question, whether in the long run the utmost satisfaction for industrial workers can ever be gained under the capitalist system, where they do not feel in any sense the owners and managers of the enterprises in which they work.

89. *Science and Civilisation in China*, vol. 4, pt 3, fig. 876, p. 262.

90. One must remember that, broadly speaking, the Chinese people had in their history neither a hereditary aristocracy nor an entrepreneurial bourgeoisie, only an intellectual élite often quite widely recruited. I record grateful thanks here to conversations with Stefan Dedijer and Ivan Divac.

91. Their contributions will appear in *Science and Civilisation in China*, vol. 7.

92. Later on, the Dignaga logic, also intensional, was brought in from India, but it never spread beyond the relatively narrow circles of those who occupied themselves with Buddhist philosophy.

93. See his 'Preface' to the *Great Instauration*, quoted in *Science and Civilisation in China*, vol. 2, p. 200; A less well known, but equally downright, statement may be found in John Webster's *Examination of Academies* (London, 1654). He wrote: 'It is clear, that Syllogising, and Logical invention are but a resumption of that which was known before, and that which we know not, Logick cannot find out; for Demonstration and the knowledge of it, is in the Teacher, not in the Learner; and therefore it serves not so much to find out Science, as to make ostentation of it being found out; not to invent it, but being invented to demonstrate and shew it to others. A Chymist when he shews me the preparation of the sulphur of Antimony, the salt of Tartar, the spirit of Vitriol, and the uses of them, he teacheth me that knowledge which I was ignorant of before, the like of which no Logick ever performed. . . . I will only conclude with that remarkable saying of the Lord Bacon: "Logick which is abused doth conduce to establish and fix errors (which are founded in vulgar notions) rather than to the inquisition of Verity, so that it is more hurtful than profitable".'

94. Actually the principle of the excluded middle is quite compatible philosophically with change, as was shown in a brilliant paper

(unfortunately still not translated into an international language by K. Ajdukiewicz, 'Zmiana i Sprzecznosc' (Change and Contradiction) first published in *Myśl Wspólczesna*, no. 8-9 (1948) p. 35; then reprinted in his *Język i Poznanie* (Language and Cognition) vol. 2 (1950) p. 90.

95. He first enunciated this 'law of inevitable succession' at the initial International Conference on Taoist Studies at Bellagio in 1968: 'Any maximum state of a variable is inherently unstable, and evokes the rise of its opposite.'

96. It is noticeable that in Europe pleasure has all too often been thought of simply as the absence of pain, rather than a positive thing in its own right, the Yang as opposed to the Yin, and both indispensable parts of living.

97. 1. Thess., 5, 21.

98. Two papers may be recommended: J. L. Cranmer-Byng, 'The Chinese Attitude towards the Natural World', *Ontario Naturalist*, 12 no. 4 (1972) p. 28; and Watanabe Masao, 'The Conception of Nature in Japanese Culture', Paper at American Association for the Advancement of Science, Washington, 1972.

99. F. S. C. Northrop, *The Meeting of East and West: an Inquiry Concerning Human Understanding* (New York: Macmillan, 1946).

100. Fêng Yu-Lan, 'Why China has no Science', *International Journal of Ethics*, 32, no. 3 (1922); see also *The Grand Titration* (London: Allen & Unwin, 1969).

101. Ch. 40, cf. Waley tr. p. 192.

102. As Chang Tsai said in his *Hsi Ming* (ca. +1066): 'That which fills the universe I regard as my own body and that which directs the universe I consider as my own nature.' All the Neo-Confucian philosophers had this nature-mysticism, this one-ning of oneself with the creative life-giving force of Nature, this love intoxication with all people and all things. 'The man of *jen*', wrote Chhêng Ming-Tao (d. +1085), 'forms one body with all things without any differentiation'. See the paper by Chhen Jung-Chieh (Chan Wing-Tsit), 'Chinese and Western Interpretations of *Jen* (Humanity, Love, Humaneness)', *Journal of Chinese Philosophy*, 2, no. 2 (1975) p. 107.

103. Mencius, II, 1, ii, 16; See also *Science and Civilisation in China* vol 4, pt 2, p. 347.

104. The classical passage on nature conservation occurs at I, 1, iii, 3. Kungsun Chhiao took the same line, see *Tso Chuan*, Duke Chao, 16th year, Couvreur tr. vol. 3, p. 272.

105. ch. 8, p. 10a; see also *Science and Civilisation in China*, vol. 4,

pt 2, p. 139.

106. *Hou Han Shu*, ch. 106, p. 13b; see also *Science and Civilisation in China*, vol. 4, pt 3, pp. 670 – 1.

107. *Novum Organum*, aphorism 129; see also *Science and Civilisation in China*, vol. 2, p. 61.

108. In words of much interest, Marco Pallis has linked this precisely with the anti-idolatry complex spoken of already. 'It may be pointed out', he said, 'that when Christianity emerged as victor from its protracted struggle with paganism, a violent reaction set in against what had come to be regarded, rightly or wrongly, as a divination of physical phenomena; a certain anti-nature bias was thereby imparted to Christian feeling and thinking that has persisted ever since.' And he goes on to show how the Renaissance gave a vast impetus to human interest in all the things of Nature while at the same time setting the stamp of profanation on them, removed all hesitations concerning what the mediaeval theologians called *turpis curiositas*, and proceeded to the wholesale ravaging of the creation on land, at sea, and in the air; from *The Sword of Gnosis*, ed. J. Needleman (Harmondsworth: Penguin, 1974) pp. 77 ff.

109. *Machina ex Deo* (MIT Press, 1969); 'Christianity is the most anthropocentric religion the world has ever seen. . . . By destroying pagan animism, Christianity made it possible to exploit Nature in a mood of indifference to the feelings of natural objects', p. 86. What a contrast to Chinese *fêng shui*, which went perhaps to the other extreme, forbidding mines, roads and industries in the interest of the blessings conferred by Nature undisturbed.

110. See also Lynn White's celebrated article, 'The Historical Roots of our Oecological Crisis', *Science*, 53 (1967) 1203.

111. Parallel currents in Israel and Islam might be found in Chassidism and Sufism.

112. These criticisms do not hold good of Eastern Orthodox Christianity to anything like the same extent. The Greek and Syrian fathers had a much more sacramental appreciation of material things than the up-and-coming West. See, for example, S. Brock, 'Word and Sacrament in the Writings of the Syrian Fathers', in *Sobornost*, 6, no. 10 (1974), 685; and the following article by V. de Waal, 'Towards a New Sacramental Theology', p. 697. Perhaps it was a true instinct for keeping intact the distinctively religious form of experience that led the Greeks to ban mechanical clocks and musical organs from their churches.

113. See the interesting book of J. C. Scott, *Health and Agriculture*

in China (London: Faber & Faber, 1952). For current composting methods in China the files of *China Reconstructs* and *China Pictorial* may be consulted. See also, for example, 'The Digestion of Night-soil for the Destruction of Parasite Ova', *Chinese Medical Journal*, 54, no. 2 (1974) p. 107, English abstracts, p. 31.

114. Even the military classics, like the *Sun Tzu Ping Fa*, counsel the 'way of weakness', and warn against driving an enemy to desperation. As the *Tao Tê Ching* says: 'Weapons are ill-omened things. All beings loathe them eternally. He who has the Tao has no concern with them.' Ch. 31, cf. Waley tr. p. 181.

115. Ch. 6, cf. Waley tr., p. 149.

CHAPTER SIX

1. Toward an Anti-Imperialist Science', by the Ciencia para El Pueblo Group (Mexico) in *Science for the People*, vol. 5, no. 5 (September 1973) p. 18.

2. H. Braverman, *Labour and Monopoly Capital* (New York: Monthly Review Press, 1974) p. 167.

3. J. D. Bernal, *Science in History*, vol. I, p. 3 (MIT Press, 1971).

4. Ibid. p. 3.

5. Ibid. p. 31.

6. See Y. Agbeyegbe and A. Habtu, *Africa before the White Man*, (Queens College – CCNY, 1971) pp. 4, 9.

7. A. Toynbee, *A Study of History* (Oxford: The Clarendon Press, 1939).

8. See W. Rodney, *How Europe Underdeveloped Africa* (London: Bogle-L'Ouverture, 1972) and his 'Problems of 3rd World Development', *Ufahamu*, vol. III, no. 2 (Autumn 1972) pp. 27 – 47.

9. Ibid. pp. 104 – 5.

10. One needs to read B. Hessen, *The Social and Economic Roots of Newton's Principia* (New York: Howard Furtig, 1971) for a more thorough Marxist analysis/example of scientific thought and research as dependent upon the political economy of the state.

11. 'Toward an Anti-Imperialist Science', p. 18.

12. Bernal, *Science in History*, p. 11.

13. R. Girling, 'Dependency, Technology and Development', in *Structures of Dependency*, ed. F. Bonilla and R. Girling (Institute of Political Studies, Stanford University, 1973) pp. 46 – 62.

14. 'Toward an Anti-Imperialist Science', p. 18.

15. Ibid. p. 50.

16. See *Science for the People* magazine for the evidence that the 'Green Revolution' feeds US multinational corporations and aids in the genocidal dependency syndrome throughout the Third World.

17. L. Fieser, *The Scientific Method* (New York: Reinhold, 1964).

18. 'Toward an Anti-Imperialist Science', p. 19.

19. Ibid.

20. R. R. 'Cable TV', in *Science for the People*, vol. VI, no. 1 (January 1974) p. 45.

21. E. Galeano, *Open Veins of Latin America*, (New York: Monthly Review Press, 1973) p. 16.

CHAPTER SEVEN

1. Clearly, I take issue here with a main trend in the work of L. Althusser and his collaborators. Let me say, however, that, historically speaking, I owe much to my reading of their (unfortunately unpublished) 'Cours de philosophie pour scientifiques' (Paris: École Normale Supérieure, 1967 – 8). Also, against this Althusserian current, I may appeal to the 1963 paper on 'Marxisme et Humanisme' by Althusser himself in *Pour Marx* (Paris: Maspero, 1971) where he analyses very clearly the persistent, though different, role of ideology in a classless society; he writes, in particular: 'le materialisme historique ne peut concevoir qu'une societe communiste elle-meme puisse jamais se passer d'ideologie, qu'il s'agisse de morale, d'art, ou de representation-du-monde' — *or of science*, I would add.

2. 'Without destroying one cannot construct. To destroy is to criticize, it is to make the revolution. In order to destroy, one must reason, and to reason is to construct. Thus it comes about that destruction carries within itself construction.' Mao Tse-tung, May 1966, on the eve of the Great Proletarian Cultural Revolution.

3. A. V. Cohen and L. N. Ivins, 'The Sophistication Factor in Science Expenditure', *Science Policy Studies*, no. 1 (London: Department of Education and Science, 1967).

4. See, for instance, P. Anderson, 'Are the Big Machines Necessary?', *New Scientist* (2 September 1971) p. 510. This extremely interesting article clearly questions the hierarchisation of science. See also the remarks of H. Casimir, President of the European Physical Society, reported in *Physics Today* (December 1972) p. 75. It might be worth noting that both Anderson and Casimir are specialists in the physics of condensed matter; the former has worked for Bell Telephone

Laboratories, the latter was once the manager of Philips Laboratories.

5. B. Brecht, *L'Achat du Duivre* (Paris: L'Arche, 1970) p. 53.

6. See the chapter on 'The Vietnam War and Physicists' in A. Jaubert and J. M. Lévy-Leblond, *(Auto) critique de la science*, (Paris: Le Seuil, 1973) p. 182; the pamphlet *Science against the People* (Berkeley: SESPA). See also professional magazines, for example *Scientia*, 107 (July –August 1972) p. 801; *Nature*, 239 (22 September 1972) p. 182; *Physics Today* (October 1972) p. 62; *Science*, 179 (2 February 1973) p. 459; *Physics Today* (April 1973) p. 11; *Science*, 180 (4 May 1973) p. 446.

7. *Le Monde* (16 October 1973).

8. A very recent illustration of this real neutralisation of criticism by the ideology of competence has appeared in an article written by D. Hafemeister, 'Science and Society Test for Physicists – The Arms Race', *American Journal of Physics*, 41, 1191 (1973). In sixteen questions, the author makes it possible for the reader to check any technical knowledge of arms-race problems. Questions and comments are relevant and instructive; and, as the author indicates, few physicists, even intimately concerned, can carry through the test correctly. The implication is clear: if you are not competent in such a subject, what right do you have to speak about it? The author has managed to introduce the problem under a purely technical and completely non-political point of view, without a word about the economic or military consequences of the answers to the questions he poses.

9. Characterised by bourgeois sociology as the 'norms' of science. See R. K. Merton, 'Science and Technology in a Democratic Order', in his *Social Structure and Social Theory* (New York: Free Press, 1957).

10. *Physics Today* (September, 1972) p. 71.

11. *Physical Review Letters*, 12, 527 (1964).

12. See H. Zuckerman, 'The Sociology of the Nobel Prizes', *Scientific American*, (5) 217 (1967) pp. 25 – 33.

13. T. S. Kuhn, *The Structure of Scientific Revolutions* (University of Chicago Press, 1962).

14. See, on this subject, M. Bunge, *Foundation of Physics* (Berlin: Springer-Verlag, 1967) and *Philosophy of Physics* (Dordrecht: Reidel, 1972); see also J. M. Lévy-Leblond, 'Quantique (Mécanique)', in *Encyclopaedia Universalis* (Paris: Le Seuil, 1971).

15. See J. M. Lévy-Leblond, 'Les inegalites de Heisenberg', *Encart Pedagogique du Bulletin de la Societe Francaise de Physique*, 7, 15 (1973).

16. W. Heisenberg, *Physics and Philosophy* (New York: Harper, 1962) p. 103.

17. J. Monod, *Le Hasard et la necessité* (Paris: Le Seuil, 1970) p. 129; English translation, *Chance and Necessity* (London: Cape, 1972).

18. M. Born, *My Life and Views* (New York: Scribner, 1958) p.148.

19. R. Feynman, R. Leighton, M. Sands, *Quantum Mechanics*, vol. III of Feynman's Lectures in Physics (New York: Addison-Wesley, 1965); E. Wichmann, *Quantum Physics*, vol. 4 of 'Berkeley Course in Physics' (New York: McGraw-Hill, 1967).

20. This idea is very close to that of Kuhn's 'paradigm'; see *The Structure of Scientific Revolutions*.

21. Galileo Galilei, *Dialogue Concerning Two New Sciences*, trans. H. Crew and A. de Salvio (New York: Macmillan, 1914).

22. Except for Jean Matricon, from whom I have borrowed the idea; see also the first pages in S. Rose, *The Chemistry of Life* (Harmondsworth: Penguin, 1966).

23. Y. Rocard, *Le signal du sourcier* (Paris: Dunod, 1962); Rocard takes great care in dealing specifically with divining flowing water, excluding as irrational other forms of dowsing (search for hidden materials or persons, and so on, whether it be with pendulums or twigs). To my knowledge the most recent attempts at serious statistical evaluations of the alleged performances of dowsers, such as published by R. A. Foulkes, 'Dowsing Experiments', *Nature*, 229 (1971) pp. 163-8, and which concluded their complete failure, did *not* deal with flowing water-divining.

24. This is the reason why many contributions collected in the *(Auto) critique de la science* come from the milieu of physicists.

25. See in particular the chapter 'Crise de la science et crise des scientifiques', in *(Auto) critique de la science*, and especially the discussions between research workers on pp. 285-8. One could also call this position the ideology of unconcern (*bof* . . . in French).

Index